Symphony of Matter and Mind

Part Six

HARMONIES OF THE MIND

Physics and Physiology of Self

Stanislav Tregub

TREGUB S.V.

Copyright © 2021 Stanislav V. Tregub

All rights reserved

No part of this book may be reproduced, or stored in a retrieval system, or transmitted in any form or by any means, electronic, mechanical, photocopying, recording, or otherwise, without express written permission of the author and publisher, except for short citations in relevant context.

For information about permission to reproduce selections from this book, please, write to symphony@stanislavtregub.com

**Symphony of Matter and Mind. Part Six.
Harmonies of the Mind. Physics and Physiology of Self.**

by Stanislav Tregub

ISBN 9798453148561

Cover design by: Stanislav Tregub

The author is not responsible for the websites to which there are links in the book, and does not guarantee that the content of these sites will remain intact and relevant to the topic.

To my sister, Svetlana

TABLE OF CONTENTS

Introduction ix
Chapter 1 Physical Approach to the Binding Problem 1
Chapter 2 Brain Polyphony and Polyrhythm 7
Chapter 3 The Emergence of Order from Complex Dynamics 38
Chapter 4 Personality as a Physical Process 59
Chapter 5 The Secret of the Unified Self 77
Chapter 6 Looking for Harmony in the Brain 100
Chapter 7 Musical Notes of the Mind 117
Chapter 8 Brain Music Notation 140
References 155

INTRODUCTION

The brain is an orchestra playing a harmonious symphony of the Mind that we experience as the unity of our picture of the world and ourselves in this world. Violations of this process, which we call mental pathologies, lead to dissonances and even complete disintegration of the picture.

How do billions of neurons perform this symphony? In other words, how does the brain create a coherent and integral model of reality while maintaining the identity of each encoded signal? In neuroscience, this question is called the binding problem.

The harmony of the Mind is a physical phenomenon, and it must be explained physically. The author solves this riddle, based on the Theory of Energy Harmony and the Teleological Transduction Theory developed in the previous volumes of the series. The book describes the physical binding mechanism that makes the symphony of the Mind harmonious and reveals the subtle nuances of its physiological implementation in the brain.

Chapter 1

Physical Approach to the Binding Problem

Had evolution not solved the binding problem, we would not be discussing it now.

Rodolpho Llinas

The number of signals entering the sensory systems and processed by the brain is enormous, and they are constantly changing. However, a normally functioning brain manages to build a coherent model of reality. It is like a jigsaw puzzle, consisting of many pieces that fit together and create a unified picture. This also concerns the model of Self that is created by the brain. Our internal feeling of personality as a single 'I' results from a combination of internal signals' representations and serves as a steady reference frame for the outside world model. In certain pathologies or under the influence of psychoactive substances that disrupt brain functioning we experience a deterioration of an integrated state of the reality model and the collapse of the unified Self. These maladaptive states show that solving the binding problem is a special function of the brain and is a key to survival.

The binding problem has two aspects:

1. The segregation problem (BP1) concerns the question of the mechanisms which allow the brain to differentiate various signals of the environment received by the sensors of our perception modalities.

2. The combination problem (BP2) is about the mechanism that integrates the representations of the signals of the outer and inner world into a coherent model of reality.

The term was coined in the nineteenth century by William James who considered the ways the unity of consciousness might be explained by a known physical mechanism and found no satisfactory answer (James, 1890). With the development of physiological knowledge about the brain, the focus shifted to functional-anatomic aspects of the problem. Trying to explain segregation and combination by anatomical structure has its grounds in the obvious spatial aspects: the brain has areas that specialize in processing modality-specific signals and various aspects of these signals, while areas further in the hierarchy are engaged in associative processing and integration. The connections between these areas start as specialized labeled lines, then converge in some regions, then diverge and converge again.

The question arises: how signals keep their identity while neural pathways converge and how the overall picture stays integrated if there are so many diverging channels? Even if information flows converge in association areas of the cortex, the binding problem remains. First, these zones themselves have numerous communication channels and consist of a huge number of neurons. Second, integration in these zones must also be combined with differentiation. Third, the convergence of flows in space cannot explain how we get a picture of the world with all its constituent details at the same time. The binding mechanism must involve both a spatial and a temporal aspect.

There is an old hypothesis that the binding problem is solved by the brain via the synchronous firing of cortical neurons at a specific frequency. Obviously, firing in unison binds in time. But how does it keep the identity of each encoded signal? This hypothesis does not solve the differentiation problem. One of the initial proponents of the theory, Christoph Von der Malsburg suggested that segregation should be supported by another mechanism without giving a hint of what it might be (Von der Malsburg, 1999). Unfortunately, the binding-by-synchrony hypothesis does not solve the combination problem too. The neuronal activity encodes signals with many parameters. How can all those parameters be combined by simultaneous firing at one frequency? This is the same as reducing a symphony to one note played by all the musicians of the orchestra at the same time. It is not a combination but a fusion. The musical analogy shows that BP1 and BP2 cannot be solved separately. It may also give us a clue about how the brain solves them.

The critics noted: "The theory is incomplete in that it describes the signature of binding without detailing how binding is computed ... Nonetheless, the theory has sparked renewed interest in the problem of binding and has provoked a great deal of important research. It has also highlighted the crucial question of neural timing and the role of time in nervous system function" (Shadlen, Movshon, 1999).

It is true that previously most of the theories were concerned only with the spatial aspects of the brain's functional-anatomic structure. But space and time are conjugate variables, and we cannot ignore one or the other. Some theories try to combine space and time in their modeling. For example, Integrated Information Theory (IIT) introduces a time- and state-dependent variable φ as a measure to characterize the capacity of the system to integrate information (Balduzzi, Tononi, 2008). The authors suggest that a network architecture that combines functional specialization with functional integration leads to high φ values. The theory has a substantial mathematical formalization but "the integration measure proposed by IIT is computationally infeasible to evaluate for large systems, growing super-exponentially with the system's information content" (Tegmark, 2016).

Still, the main problem of the model is that it does not answer the question of the physical mechanism that underlies the integrated information measure and the ability of a system to provide for specialization and integration. Thus, it does not explain the binding problem solution but only confirms that the brain solves it with varying degrees of success. The theory has been criticized for failing to answer the basic questions required of a theory of consciousness: "As long as proponents of IIT do not address these questions, they have not put a clear theory on the table that can be evaluated as true or false" (Pautz, 2019).

One more popular model, Global Workspace Theory (GWT), suggests that signals enter a specific workspace within which they spread to many sites in the cortex for parallel processing (Baars, 1997). There are detailed neuroanatomical versions of such a workspace (Dehaene et al., 2003). They are relying on the physiological fact that many cortex regions send and receive numerous projections to and from a broad variety of distant brain regions, allowing them to integrate information over space and time. Multiple sensory data can therefore converge onto a single coherent interpretation. This global interpretation is broadcast back to the global workspace creating the conditions for the emergence of a single state of consciousness, at once differentiated and integrated. However, GWT does not tackle the issue of the physical mechanism that performs differentiation and integration. It only postulates the existence of some place where the function is located.

The absence of a physical solution to the problem led to the idea that the problem does not exist. Philosopher Daniel Dennett has proposed that our sense of unified experiences is illusory and that, instead, at any one time there are "multiple drafts" of experience at multiple sites (Dennet, 1981). Some neuroscientists argue that there is in fact a disunity of consciousness in the sense that the brain processes various signals by different cell populations and this activity is not coinciding in time (Zeki, 2003). But this physiological fact does not make the binding problem non-existent. Moreover, it stresses the existence of a binding mechanism as in normal conditions we do perceive the world as a whole and not as a rotating kaleidoscope of "multiple drafts." The model of reality breaks down into pieces only in pathological states showing that some binding mechanism malfunctions. There is no way we can solve the problem by stating that it does not exist.

Some modern theories are in a controversial state. On the one hand, they claim that the problem does not exist, and on the other hand, they claim that it is somehow solved by the brain. For example, the author of the Thousand Brains Theory (TBT), Jeff Hawkins states: "The binding problem is based on the assumption that the neocortex has a single model for each object in the world. The Thousand Brains Theory flips this around and says that there are thousands of models of every object. The varied inputs to the brain aren't bound or combined into a single model." (Hawkins, 2021). So, nothing is combined — no binding problem.

But even the name of the theory speaks about the problem: how do all those thousand brains integrate into one brain? The author attempts to answer: "Voting among cortical columns solves the binding problem. It allows the brain to unite numerous types of sensory input into a single representation of what is being sensed." (Ibid). But the question of the binding mechanism remains open despite the fact that the author claims that he has closed it. Using the voting metaphor proposed by the author, we can formulate it as follows: how do votes remain individual for counting when placed in a common ballot box? This is the essence of the binding problem which has two sides: combination and segregation. They have to be solved simultaneously and there has to be a physical mechanism for that. Without an idea about the mechanism, the author has to acknowledge: "Exactly how the neocortex does this is still unclear" (Ibid).

It is impossible to answer the question of how representations integrate while retaining their identity without showing what representations are physically and how they are produced technologically by the brain. Answering the questions about the physical nature of representations would be closing the explanatory gap (Levine, 1983). Thus, the binding problem entails a set of major issues and cannot be solved separately. That is why the previous volumes of our study were devoted to covering the questions of how the brain produces representations of the signals of the world and only now we are ready to answer the question of how it binds them into a harmonious reality model.

We should get back to the initial dilemma that William James faced in the nineteenth century when he contemplated the unity of consciousness: how can it be explained physically? Previous models focus on physiology and forget that it is the embodiment of physical processes that employ a physical mechanism. Unfortunately, many neuroscientists perceive the question "What is it physically?" as the question "What is the physiology?" They try to "jump over" the explanatory gap. They are trying to understand how the binding problem is solved without formulating the problem in physical terms. The same goes for all other aspects of the Mind. They are in search of neural correlates of consciousness without even defining what consciousness is in physical terms. They are looking for a black cat in a dark room without specifying what a cat is. When they fail inevitably, the old dualistic theme resurfaces: could it be that there is a fundamental difference between the nature of the nervous activity and the nature of mental processes? Many express their disappointment by saying that if the Mind is irreducible to physiology, it cannot be explained at all. They call attempts to find an explanation "physicalism" and "reductionism," implying that it is a waste of time.

Indeed, the psyche is not reducible to physiology, but not because the psyche is something of "another world." They are just different categories. The brain is an object. The psyche is a process. We cannot reduce one to another for a simple reason: it will be a category error (basic ontological and logical fallacy). For example, we cannot reduce the flow (process) to a river (object). We cannot explain the flow simply by saying that the river is flowing. Even if we describe the river in the smallest detail down to the molecules, there will be an explanatory gap. To overcome the gap, it is necessary to understand the physical mechanisms of the flow process and how they are embodied in this particular substrate. We cannot explain consciousness by simply stating that it is created by neurons and describing their activity. No matter what details we provide, even down to molecules, description is not an explanation. Phenomenological models are useful but they do not cover the explanatory gap.

The search for correlates of consciousness means an explanation of physical mechanisms that underlie the observed physiological processes and lead to mental phenomena. Otherwise, we will remain in the old vicious circle of dualism and the explanatory gap between body and soul. True physicalism is about physics. The gap should be covered with a physical bridge in all aspects of the Mind. We started building this bridge in the previous volumes of the "Symphony of Matter and Mind" series. In this and the following ones, the construction continues. Dealing with the binding problem is part of this bridge. If we want to find an answer in physical terms, we need to posit the question in physical terms too. There is no other way.

Here is an example of how this problem is formulated in neuroscience: "In its most general form, "The Binding Problem" concerns how items that are encoded by distinct brain circuits can be combined for perception, decision, and action. In Science, something is called "a problem" when there is no plausible model for its substrate" (Feldman, 2013). This definition is quite precise but it is still about physiology resulting in mental phenomena. There is a gap even in the formulation of the problem. It is not surprising that there is still a gap in solving it.

Let's try to build a physical bridge on the level of the questions asked. What are the encoded items physically? What are the physics and technology of the encoding process? What are the brain circuits physically (not physiologically!) and what do they do technologically? These questions seem so simple, but they are the stumbling block that neuroscience has encountered in building a plausible model. Amazingly, if formulated briefly, the answers seem to be simple too.

The Mind, as the perception of the elements of the world for making decisions and actions, is the process of transducing the signals coming from these elements into their representations for forming an adequate and adaptive model of reality. Any signal is a wave of energy vibrations. From a technological point of view, the brain is a signal-processing device. Physically, neurons are oscillatory systems that encode vibrations in the energy of the world. The encoding consists of initial analysis and subsequent synthesis. The analysis is the decomposition of waves into amplitude-frequency and phase components and the determination of the contribution of different components to a given signal. Synthesis is the reverse operation of transforming the discrete measurements of various parameters into a continuous wave representation of the incoming signal. Technologically, it is an analog-discrete-analog conversion. Physically, it is the transduction of signal waves into waves of neural code and integration into a unified wave structure of the reality model.

Based on such a description of the physics and technology of the Mind, we can formulate the binding problem in physical terms: *how do waves of signal representations combine into a unified wave structure while maintaining their individual parameters?* This formulation clearly shows that segregation and combination are not only two aspects of the same process but that they may be carried out by the same physical mechanism. This mechanism is related to wave physics.

To create a plausible model of the binding process in the brain we need a plausible model of the wave processes in matter in general and a plausible model of the wave encoding process in the brain in particular. That is why the foundations of this part of the study are the Theory of Energy Harmony and the Teleological Transduction Theory developed in the previous parts of the "Symphony of Matter and Mind" project. They allow us to build a physical bridge to cover the explanatory gap between the physiological processes in the brain and the mental phenomena they create.

Chapter 2

Brain Polyphony and Polyrhythm

See deep enough, and you see musically; the heart of nature being everywhere music, if you can only reach it.

Thomas Carlyle

Throughout the previous volumes of the "Symphony of Matter and Mind" series, the guiding Ariadne's thread was the musical analogy. It helped to formulate empirically valid and testable hypotheses. It helped to understand the fundamentals of the physics of the described processes. We should continue using this thread, even if it seems to us that we, like Theseus, have already emerged from the labyrinth of the Minotaur's cave as winners. Even having answered many questions, we cannot be sure that these answers are exhaustive, and even more so, we cannot be sure that new ones will not arise.

This part of the study takes the answers to the previous questions as a foundation for dealing with further questions. The central one for this volume will be the question of how all those billions of neurons and trillions of molecules in the brain produce a unified symphony of the Mind that we experience as a coherent picture of the world and ourselves in this world.

If we take the musical analogy, we should look at the fundamental laws that allow various sounds to combine in a complex but harmoniously integral musical symphony. To understand them, we have to look at the rules that apply to musical notation. The reason the notation exists at all is that we need to keep records of the sounds in an encoded way so that a performer can reproduce them. This is analogical to the task of the brain: it has to encode signals of the world to reproduce them as the model of this world for the purpose of performing in this world.

Here it makes sense to repeat that the Teleological Transduction Theory (TTT) defines the Mind as the process of transducing signals from the external environment and the body into the internal code patterns representing these signals and constituting a model of reality for the purpose of active adaptation to this reality and maintaining the integrity of a living system. The keywords for this chapter are "code patterns."

How are sounds encoded in a musical notation and how is this encoding developed? The task of the code is to represent the continuous signals by a sequence of discrete symbols. It is the same as the written language that encodes sounds of speech by symbols of the alphabet. And just like in speech, the original method was to try to fit as much information as possible into one symbol. Drawings arose that symbolized whole melodies, as there were drawings that meant words, phrases, concepts, and phenomena. This method of recording music was found on ancient Egyptian monuments, but, most likely, it is even more ancient.

When switching to alphabets, in which symbols denoted certain sounds of speech, people tried to convey the sounds of music with letters. This was the notation in Ancient Greece, and then the Greek alphabet of music replaced the Roman one. We still designate notes with the letters A, B, C, D, E, F, G. But this coding was not enough. Why? The answer lies in the physics of sounds as waves. In it, frequency is important, but sequence and duration are equally important. The phase of the wave and its dynamics speak of the temporal structure of energy. And, of course, we must not forget about the amplitude parameter. All this cannot be conveyed by listing letters as a designation of the frequency of sound. And the letters themselves were not a convenient and compact way of encoding frequency information.

We have retained the letter designations of the notes, but we do not write them on the staff. We use symbols instead. At first, these were neumes, which were various versions of dashes, periods, commas, hooks, etc. We will not go into the history of musical code evolution in detail, although it is exciting. It is now important for us to understand the essence of the trend. Neumes originally denoted sounds and even melodic turns, the movement of a melody (voice) up and down in frequency, the character and method of performance. They did not indicate the exact frequency parameter (pitch of the sound), and they did not show the duration. The performer used them as a reference point in the music already known to him, which was still monodic (monophonic), without the simultaneous imposition of different melodies into a harmonic structure. It was flat, without depth.

But even such music had to be transmitted across distances and times. For many centuries, this happened with direct contact between the transmitter and receiver of the musical message. But gradually, the code became so information-rich that the meaning contained in it became enough to reproduce the encoded representation, and direct contact during transmission was no longer required. Yes, it was only a representation of the original signal, i.e., differences were inevitable, but the more efficient the code, the better the representation expresses the original. There is also room for the "flight" of creativity of a particular performer. But this flight cannot go to cosmic distances from the base of the original message; otherwise, it would be a completely different piece. The point of the code is to carry information about a specific signal.

Neumes conveyed relatively simple music, but they were complex symbols. These are no longer pictures for the whole melody, but still not the most effective way. Evolution went on gradually. At first, there was no information about the direction of the tune (change of frequency), intervals (ratio of sounds in pitch, frequency synchronization parameter), and durations (phase synchronization parameter).

Later the neumes were arranged in a more or less strict order along the vertical, which gave a picture of pitch and clarified the intervals. Then reference lines were added, relative to which it was possible to interpret the pitch of the neume. Gradually staves formed, where symbols of sounds were placed to give a complete picture of the melody development. Thus, a person who learned such musical notation could perform a piece that she had never heard. The code has become a very effective representation of the original.

Gradually, the stave notation came to a single standard, making it possible to transmit information over long distances in space and time. But at the same time, the information density of each neume shifted towards the temporal dimension. It was no longer a symbol of a sound of a particular pitch or a melodic move, but a discrete sign "hanging" on lines or between lines, which denoted the pitch (frequency). The symbols of the notes themselves began to indicate only their duration. This is how the mensural notation arose, where polyphonic music was notated strictly according to the rhythm (from the Latin mensurabilis, mensuratus — measured).

Alignment (synchronization) of several parts required precise frequency coding and phase matching. In monophonic music (for example, the Gregorian chant), the rhythm was determined by the meaning and form of the prayer text, prosody, and the rhythm of speech. It did not require such an exact phase coupling of different oscillations (voices, musical parts). Polyphony and polyrhythm of multi-melodic music needed a completely different level of coding to achieve the accuracy of representation and synchronization of all participants in the process. It is essential to emphasize: the accuracy and efficiency of the code created an opportunity not only for storing and transmitting information but also for integrating all members of the ensemble into a single complex polyphonic music synchronized in frequencies and phases.

The musical notation became a full-fledged spectrogram of a continuous signal with the axes of space and time: vertical lines as a frequency code; division into measures, sequential horizontal arrangement and designation of the note's duration as a rhythmic structure code. Special pause symbols have also appeared to help the performer, but in principle, the duration of the note itself is sufficient to understand the time of the pause. Line, as a spectrogram, showed the development of a given melody and rhythm.

But the music included more and more simultaneous parts, which created the depth and breadth of the range of signals that had to be reflected in the code, in the representation of the signal. The logic of the evolution of the code suggested that it was necessary to complicate the structure and hierarchy — build up the musical lines. In this hierarchy, where everyone is equal, there is a specific order:

the bottom lines are usually reserved for bass and percussion instruments as the sync baseline of low-frequency vibrations; the notes move up the lines, increasing in frequency.

The code reflected two aspects of the signal: frequency and rhythm. But there are also parameters such as amplitude and tempo. How important are they? They speak about the nuances of performance. Tempo, as the number of beats per unit of time, or beats per minute, as the "speed" of notes, indicates the music character. In the notation, it is denoted as "moderate," "lively," etc. Loudness is important for the nuance of different parts of a piece. It is also designated very vaguely: loud (forte), quiet (piano), moderately loud, very loud.

This non-specific designation is not accidental. Music played loudly or quietly, quickly or slowly, remains the same music if the essential information is preserved: the development of melodic lines and harmonic combinations (frequency aspect), rhythmic structure (temporal aspect). If these parameters change, this is a different music. As a representation of a given signal, the code should be as informative as possible in these parameters and can remain very vague about others.

The parameters of speed and volume remain in the control of the performer, conductor, arranger, sound engineer. Some composers (for example, Johann Bach) did not indicate the tempo of the piece at all. Interpretations of the tempo performed by different musicians can differ significantly. Moreover, the performer can deliberately play at various tempos and volumes depending on the desire to nuance individual parts of the piece. The specific weight of the parameters of amplitude and velocity dynamics in the music code is minimal. There is no need to spend a lot of information on them. But if the code does not transmit the melody, then there will be no music. If the harmonic or rhythmic structure of the code disintegrates, then the music will also fall apart.

Let's look at the musical code to illustrate the evolution described above. Here is an example of early cuneiform notations with a very complex hieroglyphic (ideographic) structure. You can see how each symbol carries a lot of information, and there is a wide variety of symbols:

The antique alphabetical notation:

Neumes:

Lines appear:

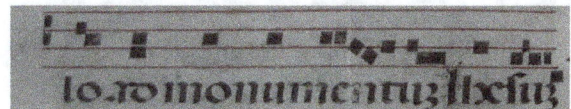

A rhythm appears with note durations:

And finally, Bach's bar notation with a reflection of melodies, harmonies and rhythms:

Fragment of Beethoven's symphony:

Modern musical notation of The Beatles song "Come together":

Please note that the symbols themselves have become quite simple, but the information richness of the code has increased by orders of magnitude. Take the bottom lines where the drum part is written: the rhythm and frequency structure is very dense (the most challenging part on the page). There is no mention of volume and tempo. And although the tempo is in principle known from the recordings of the authors' performances, any interpretation is possible. You may play it twice as slow or as fast as you can. But if you play the wrong order and the wrong length of notes, it will no longer be "Come together."

So, the modern musical code is both rich and effective. Its basic elements are no longer information-poor neumes, but also not information-overloaded hieroglyphs of the first musical notations. The set of its basic elements is limited, but a potentially infinite world of music can be created from them since they have the necessary internal parameters and can be flexibly combined with each other. They are discrete for required simplification but also continuous for sufficient complexity. Each note is simple but contains information about pitch and duration to take its proper place in the complex melodic, harmonic, and rhythmic structure of the symphony. It is separate and united with others. Such a code binds a potentially huge number of ensemble members playing their own unique part, but creating a symphony, where all sounds merge into a harmonious and unified structure while retaining their identity.

Does this sound like the requirements for neural code? Yes. Do leading models of neural code meet these requirements? No. We have already considered in detail all the technological issues of signal coding in the previous parts (see "Part Four. Algorithm of the Mind," "Part Five. Technologies of the Mind"). We also showed why the popular versions of the neural code do not meet either the technological requirements for a rich and efficient code or the reality of the observed speed and information density of processes in the brain. This is not surprising, because if the code model does not meet the above requirements, then it will not correspond to reality. Such a complex and unified system as the brain simply cannot have a code that does not reach the required level.

In short, all models of the neural code, regardless of whether they postulate that it is about the firing rate of impulses generated by neurons or in the temporal structure of the sequence of these impulses, take action potentials (spikes) for the identical discrete "shots" without any internal information. There are versions of the code that simply ignore the activity of individual neurons and assume that the information is contained only in the overall activity (population code). In any case, the basic paradigm of neuroscience is that spikes carry no individual meaning. The TTT assumes that this paradigm is wrong and has led us to a dead end. We will return to this important topic later. Now we will once again formulate the Symphonic Neural Code (SNC) hypothesis, proposed in the previous parts of the study.

Hypothesis:

The activity of a neuron has individual characteristics of the waveform (period, amplitude, phase portrait), which encodes the meaning created and transmitted by the neuron to other elements of the network. Due to this, the information density of each action potential and the resting membrane potential is very high, and the system as a whole has tremendous computing power, efficiency, and speed. All the complex and delicate logistics of the organization and kinetics of processes at the intracellular and intercellular levels are aimed at creating these parameters of the oscillatory process of each neuron.

Thus, the neural code is analogous to the musical code. Individual notes and pauses make up an activity pattern with a clear spatial and temporal organization. Integration of these patterns into the complex symphony of the Mind with its melodies (frequency patterns), rhythms (phase patterns), and harmonies (simultaneous existence of different patterns) is based on a universal physical mechanism of synchronization as frequency-phase coupling.

Such a code meets the requirements for information density, speed and efficiency. That is why this hypothesis is quite realistic. Moreover, anyone involved in the study of neural activity knows that neurons don't actually fire the same spikes. This is how they are drawn in reports, reducing all the internal parameters of the membrane potential oscillation to the peak phase and replacing real waves with phantom spikes. This makes activity analysis easier and does not contradict mainstream theories, but does not lead to deciphering the code. We need to break the impasse and return to reality.

However, such a simple theoretical assumption leads to non-trivial consequences for the practical research process. We will have to change approaches and technologies for studying neural activity. We are accustomed to counting spikes to determine their average speed. But how can we understand music if we think that it's all about the tempo of the identical notes? We are trying to place the spikes on the time axis in order to understand their rhythm. But how can we determine the rhythm if we ignore the duration of the notes themselves and the pauses between them? We measure activity in many ways to understand the meaning of the messages of neuronal ensembles. But how can we understand the musical notation of the melodies, harmonies and rhythms of these ensembles if we ignore the notes?

Let's take a look at the "music notation" that is being created by leading technologies for studying brain activity. Here is an example of an electroencephalogram (EEG):

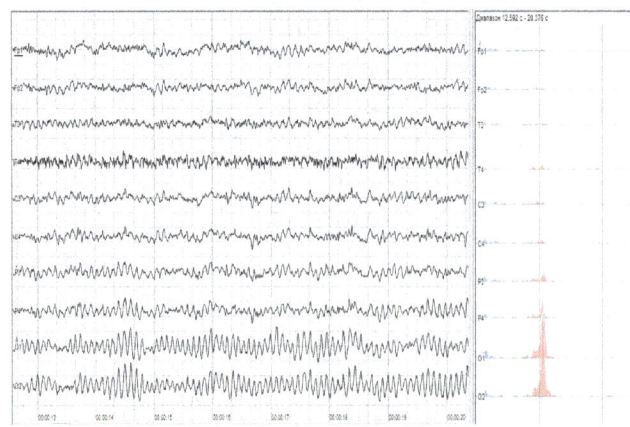

The main task of a specialist conducting an EEG study is to highlight significant features, identify their parameters and draw up a conclusion. This analysis has been carried out around the world for many decades. Now there is software that removes "artifacts," smooths out "noise," averages many records of one experiment and does other filtering operations so that a researcher can analyze the picture visually. The central meaning of this analysis is to identify the so-called "brain rhythms," examine these "rhythms" and the so-called "phenomena," i.e., areas in the signal that differ from the background picture. This analysis has diagnostic value. We will not go into the details of the methods of such diagnostics. Let's just say that it can determine some pathological conditions. For example, high-amplitude peaks are the hallmark of the epileptic activity.

It doesn't even matter that there is a confusion of concepts, and the frequency characteristics (melody, harmony) are called rhythm. And it doesn't even matter that we are talking about the surface activity of the cortex recorded by the EEG. Each technology has its limitations in spatial and temporal resolution. For now, let's emphasize one thought: with all the importance of this tool in the study of brain activity, it does not allow analyzing the neural code. It does not allow reading the music of the Mind.

These are neumes: some averaged symbols of the melodic line, which make it possible to judge the peaks and falls of the amplitude in a given range and the change in the frequency of monophonic music. But they do not help understand the details of these melodies or the intervals (frequency ratios) and do not say anything about actual rhythms, about the temporal structure of durations and sequences of sounds and pauses.

Question: is it possible to read and play music from such a musical notation? No. More precisely, you can, but if you know it in advance. Returning to the question: do we know this music in advance? No. In a circle, we return to the fact that it is impossible to read the neural code from such a record.

Now let's take the next technology. Magnetoencephalography has technological differences from EEG. MEGs are built using highly sensitive physical devices — SQUIDs (superconducting quantum interference sensor), which can measure even weak signals. Hundreds of channels and a high recording frequency allow obtaining a more detailed spatial-temporal distribution of signals. The signals recorded by MEG are less susceptible to distortion from the surface of the skull, but they are prone to distorting signals from the external environment. Isolation and filtration are required.

EEG catches both the radial components of the signal and the tangents. MEG registers only the tangents. It is sensitive to signals from a smaller fraction of the population (mainly from the cortex sulci). Still, it has better spatial resolution and allows to determine more accurately which population the researcher is observing. This, of course, makes it possible to analyze the peaks of the surface activity of the brain and diagnose some abnormalities. For example, tinnitus (auditory sensations in the absence of external signals) will manifest itself as short-term intervals of a sharp increase in amplitude and frequency in the area of the auditory cortex.

An example of a magnetoencephalogram (MEG):

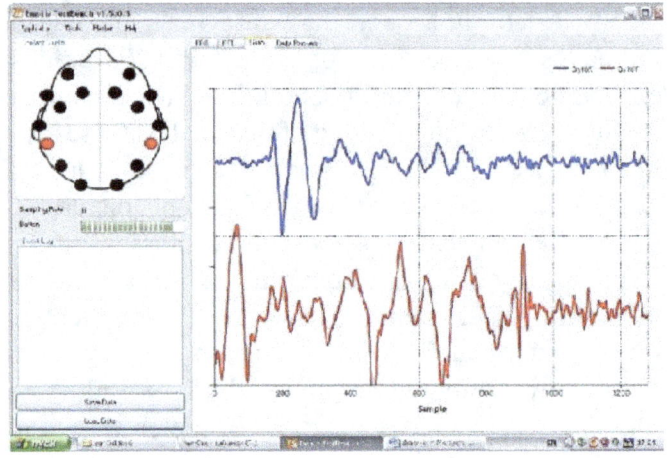

The purpose of the analysis is to detect "phenomena" against the background of general activity and then determine the location of the sources. For this, complex methods of analyzing the signal itself (including Fourier analysis) and all available technologies for determining localization (including magnetic resonance imaging) are used. But in any case, we see all the same "neumes": the picture shows either signal foci distributed over the brain surface or a graph of the amplitude and frequency of a very average signal from a population. Again, we get general hints of melodic line and harmony and no information about rhythm.

If we knew this music, we could fill in the missing data and reproduce it. But we don't know it. If we want to hear this music, we have no choice but to determine all the necessary melody, harmony, and rhythm parameters and then compose a detailed musical notation, which will give us the key to the "sounds" of the Mind. This is not an easy path, but it is easy to predict that if we do this, we will read the neural code.

An EEG and MEG graph is a spectrogram with a nominal division into frequency ranges, a mixture within the ranges themselves, the overlap of the development of the amplitude-frequency characteristics of different signals co-occurring in diverse neuronal populations. The result is a very blurry picture.

The researcher of oscillatory processes of the brain, György Buzsáki, wrote: "Because brain signals contain multiple frequency components, their relationship can be quantified using frequency domain methods ... After the signal is decomposed into sine waves, a compressed representation of the relative dominance of the various frequencies can be constructed" (Buzsáki, 2006).

But an inevitable difficulty arises: to analyze a continuous dynamic signal, an accurate spectrogram, as a combination of frequency and time domains, is required. It is impossible to mix them: these are two conjugate variables, and the principle of complementarity works. But it's entirely possible to combine them. We have been doing this for many years in other fields of knowledge. We have successfully combined both variables in music and musical notation for hundreds of years. But for some reason, in neuroscience, even if they talk about continuous oscillatory processes, the emphasis is on the frequency domain. But what does a compressed representation of the relative dominance of the various frequencies give us?

Here is an example of a frequency analysis graph of a complex sound signal:

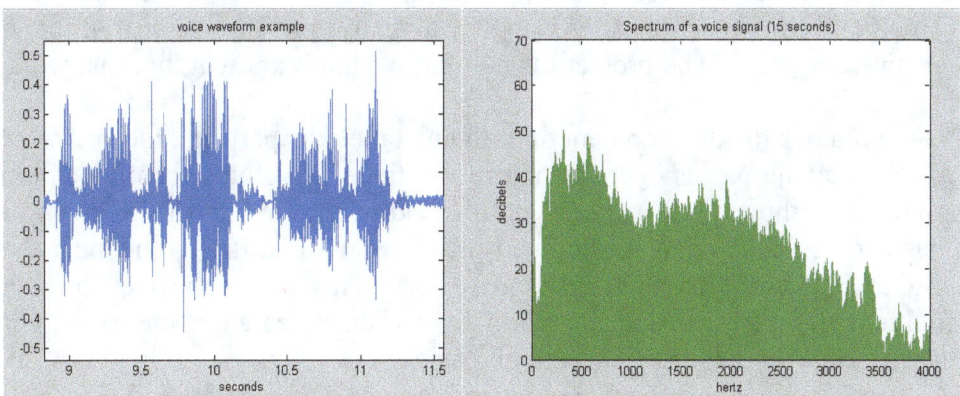

On the left, we can see the development of the signal in time; on the right, just the distribution of frequencies according to their place in the total energy of the signal. We can determine which ranges are dominant in this segment of the signal. For example, taking a piece of music, we will see that more energy was spent on instruments belonging to a specific frequency range. Neither melodic and harmonic nor rhythmic subtleties are available to us. Everything is mixed into a pile of power but with a frequency distribution. The temporal domain simply disappeared.

To make it manifest we should cut the signal into small pieces, calculate its spectral power, place them along the time axis of the signal development and obtain a spectrogram. We need a short-time Fourier transform (STFT) or wavelet transform. These are available methods in modern theory and practice of signal analysis, creating transitions from the time domain to the frequency domain and vice versa, combining both conjugate variables with varying degrees of accuracy in each of them.

A delicate balancing act is required, and the analysis of even a well-known signal (for example, music) with all its subtleties of combinations of different parts in the frequency (melody, harmony) and temporal (rhythms) aspects is a non-trivial and largely unsolved problem. It is even more challenging to analyze brain signals since we not only have a complex signal, but we do not know its meaning in advance. But since there are many difficulties, they often simplify and choose in favor of one conjugate variable in the space-time pair. The choice mainly falls on space (frequency domain). This is how both EEG and MEG signals are analyzed.

"Determining the frequency from any continuous pattern requires making measurements of time intervals before doing any calculation. However, in complex waveforms such as the EEG, where multiple frequencies are simultaneously present, it is often unclear where the intervals to be measured begin and end, that is, where the analyzed epochs should be" (Ibid).

The signal is cut into arbitrary segments, and "frequencies of interest" are selected in them. If we imagine brain signals as a polyphonic and polyrhythmic structure, where many levels develop simultaneously in time, then this approach means that we arbitrarily divide this music into bars that may not coincide at all with its internal measure. We take the most powerful ("interesting") sounds, combine them into some pattern, and try to analyze it. But the beat of this music remains a mystery. The problem is not that we don't know it, but that we do not look for it.

We remove the development of individual melodic parts in time and mix these parts, snatching pieces of frequencies from them and gluing them together in an arbitrary rhythm, which may not correspond to the rhythm of the given music. This can be called analysis, but analysis of what? And if we remember that the original EEG and MEG signal is of mixed origin itself, then the question becomes frighteningly intrusive, like a nightmare. So, the researchers are trying with all their might to escape from such an obsession.

Some simply ignore the functional importance of oscillatory and wave activity: if there is none, there are no problems. Others ignore the intricacies of polyphony and polyrhythm, simplify everything to gluing together pieces of different parts and try to hear the music. Interestingly, this is partially successful. Despite all the drawbacks of both the technologies for obtaining a signal and methods for analyzing this signal, the result can have its structure and meaning. And again, a musical analogy.

Suppose at some point the orchestra plays a relatively simple part of the symphony, where one melodic line can be distinguished, and its development in time is also simple (the rhythmic structure is prominent, since it is repeated periodically). In that case, it is quite possible even for such crude methods of analysis to make an adequate representation: the frequency spectrum has a narrow band of power density, and the bar can be estimated by periodicity. As a result, we will get a tolerable result, and it will be coherent for visual perception and assessment. If it is reproduced in sounds, we will get a rough, but close to the original version.

If we can identify patterns and trace their development in time, we can draw conclusions about the state of the interaction of different patterns, assess the degree of coherence of the process, and identify patterns. But we must remember that this is just a rough representation of either an initially simple signal or a simplified one for the convenience of analysis and that we are talking only about some melodic structures. If, as a result, we "heard" some music, this does not mean that we understood the symphony in all its complexity. It implies even less that we understood musical notation (neural code). The reason is simple: we missed the notes.

"The amplitude and waveform variability of the extracellularly recorded spike is the major cause of unit isolation errors … A further difficulty is that no independent criteria are available for the assessment of omission and commission errors of unit isolation. As a result, improvement of spike-sorting algorithms is not guided by objective measures. In the absence of quantitative criteria for unit isolation quality, interlaboratory comparison is difficult and is often a source for the controversy in data interpretation" (Ibid).

As a result, most models take the spikes for identical discrete elements and ignore their variability. For them, notes simply do not exist since it is easier to commit the mistake of inaction than to do something with the risk of an error in action. If there is no spike variability, then there is no controversy. But we cannot get away from comprehending the musical notation of the brain if we want to answer the main question about the Mind: how does it work?

This music can be complex, polyphonic and polyrhythmic, but it can have simple states. The same patterns can be found in different pieces, and there can be different patterns in the same song. The dynamics can be different, but there are patterns. We do not know musical notation yet, but we can try to determine these patterns by the development and correlation of patterns. If we grasp the basic rules of brain melodies and harmonies, then gradually, we will be able to highlight the internal elements.

Analyzing a polyphonic brain signal is not an easy task. The brain faces the same problem: it has to analyze many diverse environmental signals that co-occur in different and overlapping ranges. The brain solves this problem. Otherwise, the world would be a complete "mess" for us, and there would be no us in this world. How can our brain solve the problem of analyzing its own activity? It will solve it sooner or later, but for this, it has to do the same that it does with other signals of the environment. It is always a detailed decomposition into frequency, phase and amplitude components, and then the integration based on the existing differentiated picture. It is impossible to "swallow" the world of signals without a detailed preliminary analysis.

If the brain activity were a simple monophonic periodic signal, then there would be no problem determining its frequency response using a simple sinusoidal model algorithm and low computational costs. But brain activity can be compared to the playing of an orchestra. And this comparison is feeble: it is even difficult for us to imagine the number of musicians in the orchestra of our brain. If we take the spectrogram of even a small orchestra, made using a simple algorithm and very rough resolution characteristics, we get the following picture:

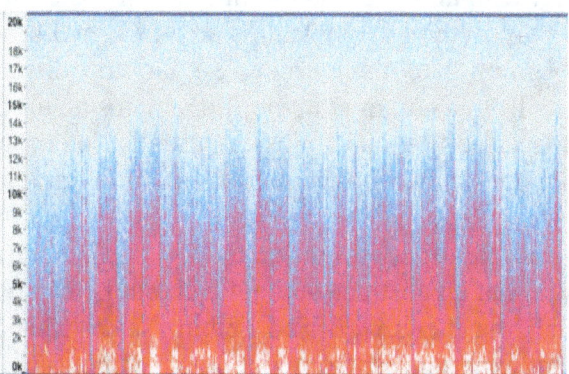

Fragment of Symphony No. 94 by Joseph Haydn

We see the frequency range, the signal strength at different frequencies with very poor differentiation, and the general picture of development over time with bursts and fading. We can say something about the melodic and harmonic structure and almost nothing about the rhythmic one. Analysis at a higher level of resolution is required. We could just as well look at the spectrogram of the sea surf noise. If we do not play the audio file together with the development of the spectrogram, then we will not believe that this is an image of a symphony.

When analyzing such a rough spectrogram, we will only see "background activity" and "phenomena" of bursts. It will tell us something about the state of the signal, but it will be very far from the music code. For example, a peak can mean different instruments playing the same tone. But we will not understand the distribution of this peak among instruments, and the meaning of this moment in the overall melodic, harmonic and rhythmic structure will remain a mystery. And we will say: here, everyone was "synchronized," and then "desynchronization" occurred. As if the rest of the time, the musicians were out of sync, just making noise, but at that moment, they were synchronized, and it was music.

It's hard to believe, but this is precisely what happens when analyzing EEG and MEG signals. This approach has led to the emergence of strange use of the term "synchronization" in neuroscience and whole theories according to which consciousness consists in these peaks of "simultaneous firing." It is not surprising that for a very long time in neuroscience, brain waves were treated as a kind of epiphenomenon, as a "motor noise": it is related to the operation of the motor; we can judge its state by it (a good mechanic can determine pathology by noise), but the motor purpose is not to produce this noise.

This attitude was largely caused by the conceptual abyss separating physiology and physics when generations of scientists simply ignored the physics of wave processes or tried unsuccessfully to find "special physical laws" for biological systems and the banal lack of accuracy of the equipment. In many ways, this tradition continues, despite the considerable breakthroughs in technology development. But the "fault" is not in engineering but in conceptual thought. So far, in many respects, wave processes remain for the mainstream a kind of "noise of the motor" without an independent role in the process of consciousness.

In engineering, with all the advances in the analysis of audio files of complex polyphonic and polyrhythmic signals, this area of signal processing technologies is still at the beginning of its journey. So far, the differentiation of various instrumental parts from a common audio file remains an unresolved problem. But if we listen to this audio file, our brains can usually determine all the instruments and their melodic lines, combine them into harmonies, and hear the details of the rhythmic structure.

If a listener is not familiar with music theory, this is not a problem. He will not be able to analyze it at an abstract-verbal level, but he can perceive all the details without even realizing it. As Gottfried Leibniz said: "Music is the pleasure that the human mind experiences from counting without being aware that it is counting" (French, 2012). But the musician, of course, is aware of what he is counting according to pitch and rhythm.

The brain will "give a head start" to any artificial signal processing technology. Its development took billions of years. But it also catches up with itself in the form of the development of engineering thought. Gradually, the engineering problem of analyzing complex signals is being solved. It is possible to single out instruments in compositions with simple and bright (by frequency response) parts. If there is a demand, there will be an offer. But in neuroscience, such a demand is just beginning to appear.

The difficulty also lies in the fact that the signal does not stand still; it develops in time. You cannot cut out a piece and say that this is the sound of the orchestra. Even one instrument in the course of the development of a part may sound differently. Moreover, even one note changes as it sounds. If you take the spectrogram of one note played on the violin, you can see that there are few harmonics and more noise at first, but then the harmonics that make up the prominent peaks of the amplitude and form the note begin to appear.

The brain is never silent while it is alive. Even in a coma, the sounding notes of the low-frequency spectrum remain. There is no beginning or end in the nervous system ensemble's current activity but various stages and phases. Consequently, the analysis will give different pictures for the same "player" or the same "part." All this complicates the research but does not make it impossible. We will have to follow the lead of our brain, which can analyze the constantly changing music of the environment. We need the concept of the approach; the rest is a technical matter (literally).

Let's take one study of the dynamics of neural interactions and the temporal correlation of brain oscillations. It was called "Long-Range Temporal Correlations and Scaling Behavior in Human Brain Oscillations" (Linkenkaer-Hansen et al., 2001). The authors begin by stating the sad situation in neuroscience concerning understanding the functions, dynamics and interactions of neural oscillations: "The human brain spontaneously generates neural oscillations with a large variability in frequency, amplitude, duration, and recurrence. Little, however, is known about the long-term spatiotemporal structure of the complex patterns of ongoing activity … The dynamic nature of these fluctuations, however, has remained unclear" (Ibid).

In other words, the brain plays complex music with a rich range of melodies, harmonies and rhythms, but this music is not yet available to us. What task did the authors set for themselves?

"We have investigated the large-scale dynamics of network oscillations in the normal human brain. To the best of our knowledge, this is the first characterization of the temporal correlations in spontaneous oscillations at time scales ranging from a few seconds to several minutes. Our results indicate that spontaneous alpha, mu, and beta oscillations have significant temporal correlations" (Ibid).

The authors tried to establish whether there is a correlation between different frequencies in an active normal brain in the main range from low to medium-high. The authors do not touch upon the physics of the synchronization process but try to find a statistical indication of temporal correlation. Here are their conclusions: "The present scaling analyses indicate that successive oscillations indeed are correlated, even over thousands of oscillation cycles ... The correlated nature of these oscillations suggests that "a burst" is only a part of a series of connected events and that the fractal structure of the signal reflects a hierarchy of bursts within bursts rather than a succession of individual or independent bursts ... One of the defining aspects of population oscillations is the ability of neural networks to establish spatiotemporal correlations with millisecond range precision and over large distances ... The exact values of the power-law exponents, on the other hand, may be related to both the biophysical mechanisms and the neural architecture underlying the oscillations ... Quantification of spatial correlations may, however, require invasive studies with greater spatial resolution than the present MEG and EEG measurements" (Ibid).

Let's try to interpret the research result. The authors found scale-invariant patterns in the correlation of different oscillation frequencies over long periods. This is called the statistical power law: the property of processes not to change their scale (functional relationship between parameters), regardless of which method or unit of measurement is applied to them. This means that the process has regularities and relations of parameters that are interdependent in specific proportions: a change in one leads to a proportional change in the other. The authors suggest that the signs of stable parameter ratios identified by them reflect some biophysical mechanisms and the structure of oscillations.

It remains to say the "magic" word synchronization, which is the physical mechanism that determines the stable scale-invariant relations of the oscillation parameters. After all, the frequency ratios' statistical signs manifest the work of the physical mechanism of synchronization, as the interaction of oscillations with different parameters.

But we have to be careful with words as they carry meanings. There are old terms "event-related synchronization (ERS) and "event-related desynchronization" (ERD). They are about short-term fluctuations in the amplitude of rhythmic activity. If the amplitude falls, it is called desynchronization, and if it grows — synchronization. This concept is used to describe the activation/deactivation of neural populations. There is a considerable amount of research on the patterns of activity of various zones caused by stimuli in terms of synchronization/desynchronization as changes in the amplitude of various frequency ranges in different states and experimental tasks.

For example: "During event-related desynchronization (ERD), spontaneous 8–13 Hz oscillations are suppressed rapidly (approximately within one cycle) by sensory stimulation, or motor activity. Furthermore, the mapping of ERD on the cortical surface has revealed transitions from spatially diffuse to focused and somato- topically specific patterns of alpha suppression, consistent with the picture of spontaneous cortical states being driven into stimulus specific configurations of correlated neural activity. We suggest that the widespread and rapid onset of ERD reflects long-range spatial correlations in the neural networks" (Ibid).

Suppose you read without a critical and careful attitude to terminology. In that case, it turns out that desynchronization leads to the emergence of a structure of patterns, correlated activity that spreads in the system over long distances and very quickly. In terms of the physical understanding of the synchronization process, this is nonsense. But a careless attitude to terms leads to such paradoxes, which contradict the physics of the process and formulations in which correlation means desynchronization. However, it is only about the fact that when the states of the brain change and when different frequencies interact, their amplitude characteristics change.

By the way, there is another term to describe the process of lowering the amplitude in low frequencies: alpha and beta blocking. It is better than desynchronization because it does not introduce confusion between categories. But it also leaves a residue of inaccuracies: complete blocking does not occur, but there is a decrease in power. A more accurate term would be attenuation, as a decrease in the power of a specific band in the system's total frequency range at a given moment in particular zones.

Why did they decide to call the decrease in the amplitude of any wave "desynchronization"? It was more than a hundred years ago, and the roots lie in the insufficient resolution of the measuring equipment of those years. At the beginning of the twentieth century, the pioneers of EEG studies (Pravdich-Neminsky, Cybulski, Beck) during experiments to stimulate afferent nerve pathways observed that active information processing by the brain was reflected in the readings as a decrease in the amplitude and even the disappearance of low frequencies. Meanwhile, the high-frequency vibrations showed a clear rhythmic pattern. Later, in the 1940s, other researchers called this state "activation of EEG" (Moruzzi, Magoun, 1949). But the term "desynchronized EEG" has already taken root and then passed on to other technologies for studying brain waves.

The pioneers of brain oscillation research had the right to call the observed phenomena whatever they wanted. But, unfortunately, the confusion in terminology is not as harmless as it might seem. In this case, it has led to "diffuse patterns" in scientific concepts, while the presence of focused and coherent patterns is desirable. Ignoring the fundamental physics of oscillatory processes and the resulting confusion in terminology continues to this day. Many people understand this, but the habit is tough to overcome.

Some even refer to their colleagues who use this terminology as "epigones" (Steriade, 1996). This is usually the contemptuous name for those who thoughtlessly imitate, uncritically reproduce other people's information with a general decrease in quality and understanding. Initially, in ancient Greek mythology, these were the sons who went on a military campaign, thoughtlessly repeating the mistakes of their fathers. Indeed, one should not repeat other people's mistakes but rather learn from them.

One can come across such paradoxical conclusions from studies: "Results substantiate desynchronization in the low-beta band as a prerequisite for fast motor performance, and add to a more complete picture of cortical-basal ganglia oscillation patterns and the anti-kinetic role of beta band synchronization" (van Wijk et al., 2017). It looks like desynchronization promotes motor skills, while synchronization interferes. But the authors of the above statement begin the article with the opposite idea: "Beta band oscillations (13–30 Hz) are a hallmark of cortical and subcortical structures that are part of the motor system. In addition to local population activity, oscillations also provide a means for synchronization of activity between regions" (Ibid).

Does synchronization have an "anti-kinetic role," or does it nevertheless link the structures of the motor zones? What kinetics can be without the connection between the elements of the system responsible for the kinetics? And what kind of synchronization are authors talking about in such different, but separated by only a few sentences, statements?

It should be borne in mind that the MEG and EEG technologies aimed at studying brain oscillations do not yet have sufficient spatial resolution to track subtle moments of amplitude, frequency, and phase modulation occurring in the neural ensemble. Even an excellent temporal resolution (around 1 msec) does not help since, in order to analyze the subtleties of the process, it is necessary to understand which element you are examining in a given temporal dynamics.

Besides, these technologies do not measure "pure" waves created by populations of neurons: the waves that MEG and EEG capture are mixed with a lot of conditionally "extraneous" noises (activity of other cells inside the skull) and really extraneous noises (external sources of waves in the surrounding space). The power of brain waves is so low (about 10^{-15} Tesla) that even the magnetic radiation of the Earth (about 5×10^{-5} Tesla) is extraneous noise. Both technologies do not penetrate deeply into the brain but receive a "picture" from the surface. Furthermore, it reflects the general, undifferentiated activity in cortex layers.

For example, a comparison of MEG readings and intracranial measurements of local potentials in vitro showed that "cortical signals measured with MEG will be an amalgam of signals from all cortical laminae in that location." But "multiple frequencies can arise from the same cortical area," and "these rhythms have independent mechanisms of generation and may, therefore, be modulated independently of one another" (Ronnqvist et al., 2013). In reality, the music of the brain is very complex and differentiated, but MEG takes a big picture of zones for a simple technological reason: "magnetic field strength decays with distance, the relative contributions of activity from superficial and deep cortical layers are unlikely to be equal" (Ibid). And we are talking not only about a wide range of differentiated activity in the layers of the cortex but also about the fact that "the extensive connectivity between the cortex and other cortical and sub-cortical structures results in the potential for great complexity in the generation of oscillations," and "it is uncertain to what extent each of the measured oscillatory signatures are inter-cortical (within cortical region) and which are intra-cortical (between regions) dependent" (Ibid).

To summarize, we can say this: with a very complex temporal and spatial structure of all elements of the system, the equipment measures an "amalgam" of simultaneously operating elements in some layers, which can create a signal of the necessary strength for registering. The technology does not differentiate the signals, and some of the signals simply do not reach its sensors (the so-called "silent currents"). Thus, a kind of "zero field" appears, about which there is no information. The earliest approaches to the analysis proceeded from simple logic: since there is no data on the process, we set a zero value for this field (minimum-norm solution). What we cannot measure is simply nonexistent.

So, the technology itself, with all its focus on brain oscillations, is not very informative in terms of the actual nuances of wave processes in the brain and does not give an idea of the intricacies of amplitude, frequency and phase modulation, without which it is impossible to create representations.

In addition to technical problems that are inevitable at this stage of development of measuring tools, there are also problems with a conceptual approach to data analysis. The absence of a piece of data is typical when analyzing any signal. Interpolation (addition of intermediate values) and extrapolation (addition outside the data interval) are always possible and necessary. But both require a model from which the interpolation and extrapolation function comes. This illustrates well the approach to the phenomenon of "event-related desynchronization" and "event-related synchronization."

How are ERD and ERS calculated? An initially arbitrary frequency range is taken (division into alpha, beta, gamma rhythms, etc., is nominal). The most reactive frequency band is determined by trial and error when performing tasks in the experiment. This signal is filtered within this band, averaged, and as a result, we get the energy of the entire band, where the subtle nuances of internal dynamics are not visible.

Of course, the analysis methods improve, but there is also a global inverse problem of any technology for measuring any process that is not directly observable: it is necessary to determine the causes of the observed phenomenon based on the measurements obtained. There are results first, and then the cause must be determined. However, the reason, or rather its interpretation, depends on the model from which the experimenter proceeds. He may pretend that he measures with an open mind. But this is a myth since projection always works, whether a person thinks about it or not. And here, the question of a concept arises. Suppose the experimenter does not look at the process from the point of view of physics and technology of signal processing by the system. In that case, such strange terms as "desynchronization," which is actually amplitude modulation, and bizarre conclusions that synchronization is "anti-kinetic" appear.

The authors of one review wrote: "Motor planning in decision making is encoded by increased gamma oscillations and decreases of alpha and beta power in the motor cortex before the execution of the motor response" (Pizzella et al., 2014). But they call this decrease in power desynchronization. A simple question arises: isn't quietly playing ensemble (be it musical or neuronal) synchronized? Does a power reduction mean the decay of frequency and phase coupling? No, it just means redistribution of energy.

The same authors wrote: "The long-range temporal correlation recovered from source-reconstructed MEG oscillatory activity correlates with both behavioral performance fluctuations and neuronal avalanche occurrences in anatomically well-identified brain regions … The interaction between the attention control system and sensory systems is served by the same frequency bands in which the sensory system's task-related response is known to occur … The putative role of phase coupling is to facilitate communication between separate neuronal populations during stimulus or cognitive processing, which may serve to regulate the integration and flow of cognitive contents on fast timescales relevant to behavior" (Ibid).

Questions arise. Isn't temporal correlation synchronization? Isn't frequency and phase coupling sync? The lack of a coherent model leads to confusion in terminology.

Suppose we look at the physics and technology of the Mind proceeding from the hypothesis that it is a process of signal transduction and creating wave representations. In that case, we can assume that amplitude fluctuations in different frequency ranges mean a simple thing: when states change, energy is redistributed from one frequency range to another. Another hypothesis is also possible: such amplitude modulation is part of the coding process. Both assumptions are physically plausible and do not contradict each other.

Many studies concerning various attributes of the Mind have revealed this modulation: "Oscillatory brain activity and its modulation have also been linked to visual attention. In particular, site-specific gamma oscillations have been linked to the individual's attention capacity, whereas suppression of theta to beta amplitudes has been correlated with attentional load. In addition to the visual modality, auditory attention has been shown to modulate alpha power in auditory regions, and specifically in relation to the information content of the auditory cue. Modulations of the power of oscillatory brain activity have also been linked to mentalizing processes such as those related to non-verbal communication" (Pizzella et al., 2014).

The authors of a review article came to the following conclusion: "Putting these experimental observations together, we suggest that a signature of engaged cortex is a reduction in the power of alpha/theta (perhaps with a lowering of frequency), modulation of gamma amplitude by the remaining alpha/theta oscillations, and an increase of the gamma duty cycle" (Lisman, Jensen, 2013).

Just before the system exhibits any activity (motor or cognitive), low frequencies decrease, and high frequencies increase their power. This pre-act modulation occurs almost instantly: within a fraction of a second before the movement of the body or thought. Analysis of the data shows that activation occurs according to the same spatio-temporal patterns during the direct execution of the movement and imagining it. It appears in an extensive network, including the higher integrators of the cortex of the frontal, temporal and parietal lobes, the premotor, motor and sensory zones of the cortex, the cerebellum, the basal ganglia, and other subcortical structures. It is a wide and coherent network, and due to its synchronization, it is capable of almost instantaneous modulation and integration of activity.

Studies show that "fluctuations over motor cortex before decision making are predictive of upcoming responses. These signal fluctuations are partly carried over from the previous response and predict a tendency to alternate between response alternatives for consecutive choices." In the context of a specific experiment, it means that "neuronal activity in sensorimotor cortex predicted which button participants eventually pressed not only after, but even before the choice-response cue, before the stimulus and more than 6s before the final motor response." The authors made the following conclusion: "As such our results accord well with a growing body of evidence suggesting that motor regions are directly involved in the process of decision making. That said, our results are also well compatible with converging data that suggest a prominent role of frontoparietal association cortices in decision making" (Pape, Siegel, 2016)

From the Perception-Apperception-Action Lemniscate (PAAL) model perspective, the experiment demonstrated the formation of a representation of motion in the higher integrators of the cortex in the projective stage of the algorithm (see "Part Four. Algorithm of the Mind"). If necessary, the system can form a representation and transmit it along the chain almost instantly, in the range of milliseconds. But if it has time to think about a choice, more integrators are involved up to the prefrontal zones.

Another study that focused on activity in the prefrontal cortex using electrocorticography (ECoG), which has much better resolution than EEG and MEG due to the direct connection of electrodes to the open brain, showed the following results. Patients undergoing preparation for brain surgery were connected to electrodes and asked to move at their own pace. Measurement of prefrontal zone activity was sufficient to classify different movements by pre-movement cues (-2 to 0 seconds). The prefrontal zone participated in the formation of the representation of the action before the start of the movement, and the time ranged from a slow period to an instant (less than a second). The most significant change in power was in the beta band (Ryun et al., 2014). In sensory networks, everything happens in a similar way to motor networks. For example, before the start of an active visual process, the amplitude of low frequencies decreases, and the main energy is concentrated in the gamma region.

First, we see the confirmation of the projection without which there is no action (neither motor nor mental). Second, amplitude modulation is an essential factor in this wave process. The change in the power of the spectrum is visible even with the simple analysis. It is much more challenging to identify the subtle frequency and phase nuances, which form the basis of the process, the subtleties of melodies and rhythms of the Mind. Unfortunately, so far, the technologies do not reach the required level of resolution. But the point is not in the technologies as such but in the conceptual approach. Engineering thought does not stand still, but it needs demand as accurate questions with correct terminology.

As the authors of one article wrote: "While there is plenty of evidence connecting the beta ERD to movement planning and selection, much less is known about the inherent dynamics ... Little is known about potential functional differences between premovement beta ERD and that occurring later (i.e., after movement onset). Likewise, how the inherent complexity of the movement modulates the beta ERD amplitude and dynamics is also not understood ... We hypothesized that an extended network of motor and association cortices would exhibit a strong beta ERD prior to and during movement, and that the amplitude of this response in several brain regions would scale with the inherent complexity of the movement sequence, thereby connecting another critical aspect of motor planning operations to the oscillatory beta ERD response ... This study is the first to demonstrate that complexity modulates the dynamics of the peri-movement beta ERD, which provides crucial new data on the functional role of this well-known oscillatory motor response. These data further suggest that execution of complex motor behavior may recruit key regions of the fronto-parietal network, in addition to traditional sensorimotor regions" (Heinrichs-Graham, Wilson 2015).

The same authors in another study found that ERD dynamics strongly depend on whether the movement is associated with a particular external signal and its parameters, informational content and temporal factors (Heinrichs-Graham, Arpin, Wilson, 2016). Let's try to draw a conclusion from these studies, based on the hypothesis about the physics of the process.

Hypothesis:

A decrease in the amplitude in the low-frequency range immediately before activity (cognitive, motor) means that the fundamental and synchronizing low frequencies go into the background when a wave of representation arises and projects, which requires an increase in energy expenditures for the formation of high-frequency components. The released energy is redistributed to the high-frequency content of the projected representation.

Possibly, subtle amplitude modulation also occurs as part of the encoding process. In any case, the change in amplitude indicates the process of creating, reproducing and projecting wave representations of movement or any cognitive acts, for which the synchronization of different wave structures is a necessary condition. That is why the phenomenon of "event-related desynchronization" (an unfortunate term for a decrease in the amplitude of oscillations) of the low-frequency range is observed immediately before the act and in areas related to the execution of the task. This "desynchronization" speaks precisely of the ongoing physical process of synchronization during the formation and transmission of representation. There is a redistribution of total energy in the system and modulation of frequency, phase and amplitude parameters to integrate and synchronize different system components participating in the wave pattern of representing a motor action or cognitive act.

Musical analogy: for mid and high-frequency instruments to play their delicate, melodically detailed parts and be heard, low-frequency instruments must fulfill their function of creating a rhythmic baseline to synchronize the entire ensemble and not drown out the rest. If we look at the physics of energy wave processes, the concept becomes so coherent that it sounds even prosaic. The system does not have infinite energy. It always works on the principle of the highest possible efficiency but with the requirement of the highest possible accuracy and speed. Only fine control over the parameters of different oscillations during the creation, reproduction and transmission of representations as wave patterns can satisfy these needs.

The fact that transitions from one frequency range to another are associated with a change in amplitude was noticed back in the 1930s. Later, when the researchers began to study EEG waves using Fourier methods for analyzing spectral power (power distribution depending on frequency), they found the following regularity of the ratio of amplitude and frequency: $A \sim 1/f^{\alpha}$. Such relationship is called "pink noise" and means that the power spectral density decreases uniformly on a logarithmic frequency scale. Thus, a wide range at high frequencies has the same power as a small range at low frequencies. This indicates the efficient use of energy in the system.

Just imagine that to increase the frequency range, you would have to increase energy costs in direct proportion to the expansion of the range. This option does not suit living systems: we do not have an endless and free resource. If we remember that most energy in neurons is spent on the operation of sodium-potassium pumps and ion channels that regulate the parameters of the membrane potential fluctuations, then it is not surprising that it is not easy for the system to provide high-frequency oscillations in a constant and global mode. The logic is simple: the higher the frequency, the more work for the pumps and channels.

In the hypothesis proposed here, the explanation for this distribution in the brain is physically plausible. If in some system there is a general distribution of energy to all oscillating elements, then an increase in the oscillation frequency within a limited energy resource naturally leads to the need to modulate the amplitude downward. It takes more energy to maintain a high frequency with a large amplitude. Fine regulation of oscillation parameters is a vital requirement. If the system cannot carry out such an adjustment, then this is reflected in the coherence of the consciousness, including the coherence of motor acts. It is no coincidence that studies of changes in the amplitude of low frequencies in patients with impaired motor activity show dynamics that differ from normal.

Let's take one example: the article on dysregulation of rhythms in patients with Parkinson's disease (Heinrichs-Graham et al., 2014). This disease is one of the most striking pathological conditions of the motor system, manifested in external symptoms in the form of muscle rigidity, difficulties in performing voluntary movements, general hypokinesia, tremor, and instability of body position. Usually, Parkinson's disease (PD) is considered a pathology in the subcortical zones, particularly degenerative changes in the substantia nigra, basal nuclei, and disorders in the production of the activating neurotransmitter dopamine. New research has shown that the regulation of the entire cortical-subcortical motor chain is impaired.

The authors note: "A widely replicated finding in patients undergoing DBS surgery is that the subthalamic nucleus (STN) exhibits pathological beta synchronization, which eventually entrains the entire basal ganglia–cortical network and thereby serves to block volitional movement … Patients with PD exhibited significantly diminished beta desynchronization compared with controls prior to and during movement, which paralleled reduced alpha desynchronization … Patients have significant difficulty suppressing cortical beta synchronization during movement planning, which may contribute to their diminished movement capacities … Alpha desynchronization prior to movement is linked to premovement beta desynchronization, so it is probable that such beta dysfunction carries over to alpha activity. Potentially, a certain degree of beta desynchronization is necessary before neurons in this cortical area synchronize at the faster gamma firing rate that initiates movement execution, just as alpha desynchronization in visual cortices precedes active visual processing in the gamma range" (Ibid).

The authors use the term synch/desynch in the sense of change in the amplitude of waves of a specific range. Now let's read this result in the light of the above hypothesis: patients showed no decrease in the amplitude of low frequencies (alpha and beta range) during the formation of the representation of movement, which disturbed the energy balance of the general process and led to the impossibility of creating a coherent representation in the entire frequency range, which is required for the full and smooth movement process.

Musical analogy: drums and bass instruments drowned out the whole orchestra; violins could not hear themselves and did not have enough energy. The ensemble collapsed; the music of the movement was not created. Again, we have to emphasize: the patients showed not "diminished desynchronization" but diminished synchronization if we use the term in its actual physical meaning. Perhaps those who use the terms ERD and ERS think that loud bass (low frequency) instruments are in sync and quiet ones are not. I dare to disappoint them: if they are playing music and not making noise, then they are synchronized in any case, loud or not. For ensemble play, dynamics control is critical, since low frequencies can drown the rest.

This is the physics of wave processes. In an ensemble with a common energy resource (the brain is such an ensemble), such a condition is pathological. And if we consider that in a natural environment without the help of society living organisms with such motor impairments as in PD do not survive, then it is fatally dangerous.

The authors write that "patients have significant difficulty suppressing cortical beta synchronization during movement planning, which may contribute to their diminished movement capacities" (Ibid). The translation is required again: they have significant difficulties modulating the amplitude of low-frequency oscillations, which violates the synchronization process across the entire frequency spectrum. It is a "pathological synchronization" of desynchronization. Unraveling the jumble of confusing terminology and conceptual mess is not an easy task.

By the way, difficulties with modulating the amplitude of different frequencies are also observed in children. They have poor return to high amplitude low frequencies after movement. In the developing system, adequate regulatory mechanisms have not yet been established, including modulation of the neural ensemble into a coherent synchronized state.

Typically, during movement, there is also increased gamma activity. "Unlike the beta response, the gamma response reaches maximum amplitude about 100ms after movement onset and is very brief, lasting only 100–200 ms, at which time it quickly dissipates … The gamma response is also more powerful during the first movement of a repetitive sequence than in succeeding movements" (Ibid).

Again, we consider the results in light of the above hypothesis. To reproduce a wave representation containing a wide range of frequencies, it is necessary to redistribute the power from the "bass line" into a spectrum of higher frequencies with all the nuances of this representation, all the melodies and rhythms of a given body or thought movement. The more complex the motor and cognitive act, the greater the range involved. But the larger the frequency range, the higher the energy consumption. The system redistributes since it has neither extra energy nor the need to maintain low frequencies at a high-power level when the time for the nuances of high frequencies comes.

When the movements repeat, there is no need to use the high-frequency range at full power: the representation is already stable, and "everyone is heard." But as soon as the act is completed, the representation, as a wave structure of a wide range, disintegrates: the high frequencies simply subside. If the reproduction of the representation is required, again, there is a substantial decrease in amplitude in both beta and alpha, and high-frequency instruments begin to "play their part," creating melodically subtle structures. It is the ordinary course of events.

As the authors note, "healthy controls exhibited a well-established pattern of oscillatory neural activity before, during, and after movement onsets in brain areas associated with motor processing ... If patients with PD experience aberrant beta synchronization, it is possible that they are unable to "break through" this beta synchronization in order to initiate the premovement beta ERD necessary for proper movement" (Ibid).

And again, translation into the terms of the physics of the process is required. Patients have a too high power of low frequencies, and a violation of the process of their attenuation does not allow redistribution of energy into the high-frequency spectrum for the formation of a complete and synchronized wave representation of movement. Everything falls into place if we proceed from the hypothesis about the wave nature of representations that require broadband frequency range to form a complex structure and synchronization of different frequencies that requires fine regulation of wave parameters (frequency, phase and amplitude).

What hypotheses do the authors mention? There is an "idling hypothesis," which says that the motor zones are highly synchronized during rest but desynchronized during movement and again return to the primary synchronization state after it. Another hypothesis says that the jump in the amplitude of low frequencies after movement is an inhibitory mechanism. The authors note that the second hypothesis may partially explain why PD patients cannot stop movement and lose fine motor control. But the problem is that patients cannot stop tremor and start a normal movement. There is a violation of general regulation and not just inhibition.

Interestingly, both hypotheses are partially correct. Low frequencies play both an inhibitory role when they take energy from the high-frequency range after completing the movement and an activating function when they provide a basis for bringing all the elements involved in forming the wave of a given representation into a single whole of a synchronized structure. However, the above hypothesis within the TTT is a hypothesis of a more comprehensive class. It covers these hypotheses but with an explanation of those phenomena that the previous ones cannot explain. Failure in the regulation of oscillation parameters due to impaired dopamine function in PD leads to disruption of amplitude modulation. Low-frequency oscillations, instead of serving as a basis for synchronization, "drown out" the rest of the ensemble members. This desynchronization of neuronal populations involved in motor representations leads to externally observable motor dysfunction.

We all have persistent muscle tremor in the 8-12 Hz range, but they are so low in amplitude that we don't notice them. We have already addressed this issue in the previous part of the study. It is worth repeating here that this basic muscle pulse is in the same frequency range as the neural one. It does not depend on the parameters of movements and is always active as a pacemaker. But with PD, its amplitude significantly exceeds the norm, producing a pronounced tremor of the limbs, head, other parts of the body, and even the whole body. As the disease progresses, the amplitude of the tremor increases. This is how the violation of harmony in the brain ensemble manifests itself.

Low frequencies cannot be called "idle." In general, the analogy with the engine, its RPM, and shifting from one gear to another is intuitive. But there is one subtle point that makes it more of a metaphor than an analogy. In a real car engine, there is only one total RPM. It can change, be lowered at idle, increase when shifting gears, return to a low one in a higher gear, etc. When this analogy is applied to the brain's work, it seems that one motor is running, and it has one frequency at a time. It is changing, but it is the only one. This analogy is misleading because there is a wide range of frequencies in the brain working simultaneously. It is an ensemble of "motors," or rather, an ensemble of musical instruments with different frequency characteristics.

The idling hypothesis has been refuted by many studies. For example, alpha activity increases with memory demands (Jensen et al., 1999; Klimesch et al., 1999; Park et al., 2011; Scheeringa et al., 2009; Tuladhar et al., 2007). The alpha activity in the primary sensorimotor cortex decreases contralaterally to the stimulated hands and increases ipsilaterally with the increase being the best predictor of performance (Haegens et al., 2010). This increase may serve as an active inhibition of task-irrelevant but potentially interfering activity. Studies show that slow delta oscillations control gamma activity (Handel, Haarmeier, 2009). Theta oscillations modulate higher frequency bands during cognitive, memory, spatial orientation and movement tasks (Bragin et al., 1995; Belluscio et al., 2012; Colgin et al., 2009; Canolty et al., 2006; Canolty, Knight, 2010). Studies of higher-level cognitive functions in both humans and monkeys show beta increases in prefrontal regions during decision and updating (Donner et al., 2009; Haegens et al., 2011, Spitzer et al., 2010). Studies of human brain activity during combined visuomotor tasks found long-range task-related coupling between the primary motor cortex and multiple brain regions in the low-frequency band (Jerbi et al., 2006).

The list of studies can be continued but these are enough to show that low frequencies do not disappear anywhere and continue to play their part. Any representation, be it sensory, motor, or the highest flight of our thought, begins with this basic rhythmic structure on which any rhythmic and melodic nuances, any improvisation of the music of the Mind can be superimposed. The brain is an orchestra with the distribution of frequency and rhythmic parts, in which basic low-frequency ones create context (from Latin contextus — connection) and high-frequency ones create content. The brain is a physical device that uses fundamental laws of wave physics.

Here is an illustration. In one experiment the EEG recordings were taken while subjects had to perceive the same objects in three different modalities: as spoken words, as written words and as pictures. The authors analyzed the electrophysiological pattern underlying all three ways of presentation by computing spectral coherence of inter-areal cross-correlation in six different frequency ranges. By coherence they meant a statistical measure of repeated correlations between events in the frequency domain with a stable phase relationship. Thus, statistical coherence helps to unravel frequency and phase interactions as physical coherence.

Here is what the authors noted: "While local synchronization during visual processing evolved in the gamma frequency range, synchronization between neighboring temporal and parietal cortex during multimodal semantic processing evolved in a lower, the beta1 (12-18 Hz) frequency range, and long range fronto-parietal interactions during working memory retention and mental imagery evolved in the theta and alpha (4-8 Hz, 8-12 Hz) frequency range. Thus, a relationship seems to exist between the extent of functional integration and the synchronization-frequency … Synchronous activity in the gamma range was repeatedly shown to decline with distance … Simulation studies have shown that gamma synchrony is lost over longer distances and that lower frequency interactions are better suited to sustain long range synchronization … Summarizing, we propose that long range integration during processes extending over more than one area is mediated by correlated activity in lower frequency ranges" (von Stein, Sarnthein, 2000).

These physiological facts can be explained by the physics of wave propagation. For an intuitive description, let's take sound waves as an example. If you have speakers with good frequency responses so that you can hear different frequency spectra in a sufficiently differentiated manner, then you can perform a simple experiment. Turn the music on and walk around the room. You will immediately notice that a change in position relative to the front of the speakers leads to very significant changes in the perceived frequency range. The disappearance of the high frequencies is especially noticeable when you move sideways from the front. We can think of this process as a cone spreading from an oscillating source of the wave: the higher the frequency, the narrower the cone. Low frequencies travel well in all directions. They have an advantage inherent in their essence: they are long-wave. The subwoofer can stand anywhere in the room, even under a table or other obstacle, and pump its bass well. But the high-frequency buzzer must be placed very precisely and without interference; otherwise, it will not perform its function.

Distance also affects. If you leave the room, then as you move away, the audible range will be less and less filled with mid and high range until only the bass remains. Here the absorption coefficient plays a role, which in a homogeneous medium is proportional to the square of the frequency. If the frequency of the wave is increased by ten times, the absorption coefficient will increase by a hundred times. In an inhomogeneous medium, this process is not so linear, but the pattern is preserved. All these factors are taken into account when designing acoustic rooms.

We can consider the brain as a musical hall, where the patterns of wave propagation are taken into account to the smallest detail. As we showed in the previous part of research, the brain has to apply various technological solutions to take advantage of the waves as a means of communication and compensate for the disadvantages. The advantage of each range can be a disadvantage depending on the functional point of view. Low frequencies are long-range, and high frequencies are local. But low frequencies are scattered, and high frequencies allow concentrating energy in the right direction and in a specific place. High frequencies are specialists of a narrow profile, and low frequencies are specialists of a wide profile.

The lows have another important feature: they act as carriers. The answer lies in the physics of waves and synchronization. The phase space of the lows unfolds more slowly; many phases of fast oscillations can fit into it and quickly lock with the low ones. But here, too, the advantage turns into a disadvantage (it all depends on the functional point of view). Low frequencies create much less content as they fit fewer patterns per unit of time. That is why the brain uses mostly high frequency spectrum to create the content. The physics of the sync process allows the superposition of such content onto the context and the extraction of content from the context very quickly and efficiently while maintaining targeted and differentiated routing (multiplexing and demultiplexing).

As we have discussed in the previous volume, the convergence of neural pathways at the higher level of processing in the cortex presents a conundrum for the standard notion of information flows in the brain being train spikes traveling through neurons as wires. How can the brain differentiate representations if they go through the same channel at the same time and consist of various spike trains? How do those trains not get into a traffic jam? How do they retain identity if everything is mixed up on one track? Parallel processing does not solve the problem as there are simply not enough tracks. For example, a long time ago it was noted that bundles of white matter connecting the primary visual cortex V1 with other visual areas are clearly insufficient in volume for the simultaneous transmission of the state of all neurons in this area (Pitts, McCulloch, 1947). This also applies to other cortex areas and other modalities of perception.

The higher-order integrators of the cortex use composite technology where the encoded parameters are transmitted over common channels. This simplifies the connection between the modules of the system, which is highly complex anyway due to the vast number of network elements. The composite version saves space, time and energy during transmission. But it means that the brain has to deal with the problem of the transmission of multiple data streams over one physical communication channel. This requires multiplexing technologies that include frequency division, wavelength division, time division, and on-demand multiplexing. We offered the hypotheses about the transmission technologies of the brain that allow for multiplexing within the same anatomical pathways (see "Part five. Technologies of the Mind"). It is all about the physics of the Symphonic Neural Code (SNC) that is not about trains of identical discrete spikes but about waves of continuous oscillations. It is all about transmission via white matter not as a bundle of wires but as a network of waveguides. Wave physics allows for different patterns to go by the same channel at the same time. Receivers need to be tuned to the frequency and phase parameters of waves that represent the encoded signals. For combining all those parameters and differentiating them the universal mechanism of frequency-phase coupling is used.

Let's bridge this technological and physical description with physiological reality by taking the example of our visual system. In the primary visual cortex V1 (initial integrator) the neurons encode incoming signals that have been transduced by receptors and modulators in the preceding parts of the visual tract. The data streams are going via component channels meaning that different neurons of V1 process different signals or various parameters of the same signal. These neurons have converging connections with postsynaptic neurons in further parts of the cortex responsible for encoding the final representation of the signal (intermediate integrators). But all these representations converge into the model of the whole visual field at the level of higher integrators of the cortex. The question is how does the brain choose between irrelevant and behaviorally important signals? We need to note that this happens in a matter of milliseconds and sometimes even faster. So, the mechanism must be specific, flexible and fast. The only mechanism that is physically and technologically up to these requirements is synchronization of oscillations. As we proposed, higher frequencies create content of the Mind, so we can expect that the processing of behaviorally relevant stimuli will be reflected in these ranges.

In one experiment, a group of researchers developed a high-density ECoG grid of electrodes and implanted it subdurally onto the brains of two macaque monkeys to obtain simultaneous recordings from 252 electrodes across the V1 and V4 areas of visual cortex (Bosman et al., 2012). The monkeys were presented with two stimuli either separately or simultaneously. One of the stimuli was behaviorally relevant, the other was irrelevant, and the monkey's behavior indicated that the relevant stimulus was attended to and the irrelevant one ignored. At the level of V1 the stimuli activated separate groups of neurons, while in V4 there were groups activated by both stimuli to approximately the same degree. When presented separately they induced a gamma range oscillation in the respective V1 group which entrained the V4 group. When presented simultaneously, both stimuli induced gamma oscillations in their respective V1 group. Crucially, when one of the two simultaneously presented stimuli was attended, only the corresponding V1 gamma managed to entrain the V4 gamma. The ignored stimulus induced a gamma rhythm in V1 but failed to entrain the V4 gamma. In summary, the experiment shows that V4 sites synchronize selectively with those V1 sites that are activated by the behaviorally relevant stimulus. The authors wrote: "Frequency bands of gamma activities showed substantial overlap containing the band of interareal coherence. The relevant V1 site had its gamma peak frequency 2-3 Hz higher than the irrelevant V1 site and 4-6 Hz higher than V4. Gamma-mediated interareal influences were predominantly directed from V1 to V4. We propose that selective synchronization renders relevant input effective, thereby modulating effective connectivity" (Ibid).

Even within one range there are various frequency levels that allow for directed channeling of information flows within the same anatomical channel. There are also complex phase coupling configurations that create rhythmic scheduling of the data streams. Studies show that gamma-band coupling does not occur at zero phase difference but has variations of phase shift (Bastos et al., 2015; Grothe et al., 2012; Jia et al., 2013; Zandvakili, Kohn, 2015).

There has been plenty of evidence that the phenomenal experience of a specific mental image is correlated with the synchronization in the gamma band. Especially of interest are experiments when the object representation is created out of ambiguous stimuli (so-called 'perceptual illusions'), so that it is possible to measure the brain activity at the moment of the mental image creation reported by the subject. For example, a team of researchers showed that the perception of the Kanizsa triangle corresponds to phase-locked gamma-band activity (Tallon-Baudry et al., 1996). The same team compared the gamma-band responses between pictures that are meaningless to subjects and the identical pictures after subjects were trained to perceive a hidden dalmatian dog. The result also revealed enhanced gamma-band activity associated with the perception of the dog (Tallon-Baudry et al., 1997). They found sustained gamma activity during a task that required keeping a certain image in memory (Tallon-Baudry et al., 1998).

Synchronization and interaction of brain areas in the gamma range has been studied for decades (Engel et al., 1991a; Engel et al., 1991b; von Stein et al., 2000; Fell et al., 2001; Schoffelen et al., 2005; Buschman, Miller, 2007; Womelsdorf et al., 2007; Siegel et al., 2008; Gregoriou et al., 2009; Colgin et al., 2009; Popescu et al., 2009; Sigurdsson et al., 2010; Hipp et al., 2011). Overall, there is so much accumulated data concerning the gamma-band oscillations during mental activity that it led to the hypothesis that this range is "the neuronal basis for consciousness" (Llinas et al., 1998). We will consider this hypothesis in detail in the following chapters. Here we need to stress the main point within this chapter: the Mind is not about synchronization in a specific frequency range; it is about cross-frequency coupling at many levels of the oscillatory activity of the brain.

In an article called "Rhythms for Cognition: Communication through Coherence," Pascal Fries analyzed data from numerous studies and came to the conclusion: "The experimental evidence presented and the considerations discussed so far suggest that top-down attentional influences are mediated by beta-band synchronization, that the selective communication of the attended stimulus is implemented by gamma-band synchronization, and that gamma is rhythmically reset by a 4 Hz theta rhythm ... Visual scenes induce many local gamma rhythms with varying strength and frequency, reflecting the bottom-up stimulus salience and stimulus history. The resulting gamma landscape in e.g. V1 thus reflects stimulus properties, experience and top-down influences. At a given time point, one out of these coexisting gamma rhythms succeeds in entraining postsynaptic neuronal groups. This gamma entrainment allows to transmit a stimulus representation and to selfishly shut out competing stimuli's representations ... Thus, several rhythms and their interplay render neuronal communication effective, precise and selective ... Different rhythms coexist and are often synchronized to each other or nested into each other ... Patterns of synchronization change dynamically with stimulation and behavioral context in a way that strongly suggests that selective coherence implements selective communication" (Fries, 2015).

What the author is describing is actually a functional role of various frequencies in the process of encoding, transmission, storage and decoding of representations in the bidirectional iterative PAAL algorithm. The reason Fries is focused on the connectivity function (he even calls his model Communication Through Coherence Theory) is that his view on the neural code is within the standard firing rate paradigm. Within the Symphonic Neural Code hypothesis of TTT oscillations at various frequencies and with different phase portraits not only serve communicative role but are the physical mechanism of creating representations as wave patterns.

The authors of one study wrote: "Whilst there is extensive knowledge about the physiological mechanisms responsible for different frequency components, not much is known about the cellular and network mechanisms of the interactions between these components. Indeed, in order to validate the hypothesis of cross-frequency coupling in the brain, a biophysical theory needs to be put forward as to how a neuron or an ensemble of neurons physically implements the coupling" (Deco et al., 2017).

Exactly, to cover the explanatory gap we need to bridge the observed physiology of the brain and the resulting mental phenomena with a model of a physical mechanism. The Teleological Transduction Theory developed in the previous parts of the study is an attempt to create such a biophysical theory. We looked at the details of the encoding, transmission, storage and decoding processes in the brain that result in the creation of representations and the overall model of reality. In this part we will look at how this ensemble with vast number of participants and multiple structural levels creates the coherent polyphonic and polyrhythmic symphony that we call the Mind.

Chapter 3

The Emergence of Order from Complex Dynamics

The same organizing forces that have shaped nature in all her forms are also responsible for the structure of our minds.

Werner Heisenberg

The engineer Jeff Hawkins recalled that he was struck by the confessions of the Nobel Prize laureate Francis Crick that neuroscientists have a lot of data but no "ideological framework." Hawkins noted that it is the same as saying, that we have no idea how the brain works. Crick was like a boy who said: "The emperor is naked" (Hawkins, 2005).

It remains to add that this remains true today. A framework is necessary; otherwise, we simply drown in an ocean of data about the brain. With the framework in place, the data as if by "magic" fall into place, and there is an understanding where there are gaps that have to be filled.

The framework for the Teleological Transduction Theory (TTT) and the Perception-Apperception-Action Lemniscate (PAAL) model is the Theory of Energy Harmony (TEH). The central hypothesis, as the foundation for the entire framework, is that all processes in matter-energy are based on the synchronization of wave structures during the creation of stable systems and desynchronization during their decay. Oscillations of energy with various amplitude-frequency characteristics and phase trajectory in case of harmonic frequency and phase coupling within the synchronization region create forms of matter. When they leave the sync region, they disintegrate, freeing up energy flows for new configurations. Transitions and transformations of energy, changes in the characteristics of its oscillations create all the variety of phenomena that we call matter. Synchronization is a universal mechanism of energy interactions and the creation of order from complex dynamics.

A further primary hypothesis, which sounds almost like an axiom, is the assumption that living organisms are part of this world, material and, therefore, use the same physical mechanisms for their formation, existence and preservation of integrity. Birth and life are acts of harmony; death is the disintegration of harmony.

The mechanism of interaction in inanimate and living matter is fundamentally the same. Any living organism is a self-organizing system of interacting and energy-exchanging self-sustained oscillators with various amplitude-frequency characteristics and phase portraits. The internal interaction of the elements of the biological system, generating different frequency modes and phase trajectories, is based on the creation of a network of feedforward and feedback links for mutual adjustment of parameters, fine frequency-phase tuning, and bringing the entire ensemble to a single harmony. An adaptive state is a consequence of internal harmony-synchronization, and a maladaptive pathological condition means a violation of synchronization and disharmony.

A further hypothesis, which logically continues the previous ones, is that the nervous system in general and the brain, in particular, is also such an oscillatory complex dynamic system. As environmental signal processing and creating representations for orientation in the environment and actions, the Mind cannot occur without a universal physical mechanism of synchronization.

The orchestra of the brain plays a unified, coherent symphony consisting of many different parts. The sequence and duration (rhythmic pattern) of activation of various members of the brain ensemble produces a unique temporal structure of each representation, which can combine with other similarly unique forms and create the polyrhythmic depth. Each pattern's frequency characteristics (melodic structure) can be coordinated with other patterns, thus forming polyphony and harmonic depth of the Mind. Here again, we should emphasize that synchronization means frequency and phase coupling of oscillations with different parameters. This complex phenomenon cannot be reduced to a complete unison of fluctuations or precise timing of discrete events. We will come back to this question later, as it remains one of the most confusing in studying the brain's work.

As one of the veterans of neuroscience, Marcel Verzeano, wrote: "Few words in the vocabulary of neurophysiology have been more misused than "synchronization" (Verzeano, 1972). Since then, this misuse has been on the rise. In the 1950s, Verzeano showed that neural activity reflected on the EEG in the form of waves is sequential and not synchronous in the sense of simultaneity and summation of action potentials. He recognized the oddities in the use of the term "synchronization," but with all the reservations continued to use it in his articles for a simple reason: "So far, neurophysiologists and electroencephalographers have not agreed on a better term" (Ibid).

Traditions, even strange ones, are very tenacious. But this is a pathological state of the model fixation without proper adjustment.

Verzeano wrote: "Investigations based on the use of recordings obtained, simultaneously, with electrodes or microelectrodes have indicated that synchronization is a dynamic process characterized by a highly organized circulation of activity through the neural networks of the cortex and of the thalamus. The detailed analysis of this circulation ... has shown that the degree of rhythmicity (or the level of synchronization) of gross waves corresponds to the degree of rhythmicity in the circulation of neuronal impulses, to the velocity of circulation, and to the number and level of the neurons involved in it ..., the pathway of circulation extends along a series of loops distributed through the neuronal networks" (Ibid).

It is not a linear summation of postsynaptic action potentials. He called for this concept to be revised entirely.

The fact is that the pioneers of EEG studies initially thought that the visually observed waves of instrument readings reflected the simultaneous discharges of neurons and the sum of their action potentials in the cerebral cortex. But this, to put it mildly, is not entirely true: the EEG reflects the average signal of all energy flows that reach the electrodes on the surface of the head. Wave activity recorded by voltage change sensors (voltmeters) shows changes in the energy level of many processes in and around neurons. The EEG signal is averaging both in space and in time.

EEG and MEG do not differentiate either discrete events or nuances of the continual process and its phase both in the context of one neuron and the entire population over which the sensor is placed. They show the general change of flows in neurons and outside them, the frequency of this change and the amplitude. Individual neurons do not produce "sounds" in every cycle of the equipment, and their single spike is too "quiet" for the sensor.

The wave on the monitor screen is a wave of a density of activity of some cells in a particular area close to the surface of the skull near sensors, plus all possible extracellular flows and changes in their potentials. And all these streams must also be narrowly focused in a specific direction in space for the sensor to register them. We cannot compile a full musical notation based on such data. Individual notes, melodies and harmonies are not just mixed in a heap, but chunks are snatched from this mixture and mixed into another heap.

As one of the leading EEG researchers, Walter Freeman III, wrote: "Brain potentials (EEG waves) appear to have somewhat the relation to wave activity of neural masses that flow patterns have to temperature and pressure waves in atmospheric storms. They are observable side effects that are of interest mainly because they give access to the internal dynamics" (Freeman, 1972).

Yes, they provide some insight into the internal dynamics and are very useful for diagnosing certain states of the system. But until we understand what waves underlie the observed "side effects," this "weather" will remain a mystery to us. If we do not measure the parameters that lead to "flow patterns," we can hardly be called meteorologists. We can look at the sky and determine the general state of the weather from it, and experienced people (for example, sailors) can determine dynamics by this state, make diagnoses and even forecasts. But this does not mean that they are reading the weather code, which consists of many parameters of oscillatory energy processes. That is why, when going to sea, they rely not only on their forecast based on the current general state of the sky. They also listen to the weather forecast of meteorologists, who, with all restrictions, have much more in-depth information about the parameters leading to current and future "side effects." The weather analogy also works since the physics of the oscillatory process is the same: be it the music of sounds, the music of the weather, or the music of the brain.

By the way, Freeman's article was called "Waves, Pulses, and the Theory of Neural Masses." Back in the 1970s, he shrewdly wrote about the pulse-wave dualism in the work of the brain and emphasized that "the EEG contains little or no information in itself. Neither waves nor pulses can be read without the other" (Ibid). Of course, we cannot read the music of the Mind if we do not have information about the notes and their place in melodies and harmonies.

In the preface to a collection of articles spanning several decades of work with the intriguing title "A novel pathway into brain dynamics," Freeman wrote: "As a dynamical system seen from the outside, the brain takes inputs in the form of stimuli and gives outputs in the form of responses. We should not begin with the whole but with the smallest part that will suffice. For many purposes that is the neuron … From this simple beginning, which forms a main foundation of modern cellular neurophysiology and more recently of the new field of neural networks, we generalize to arrays of interconnected neurons and to trains of impulses in an unlimited variety of network configurations. Herein lies a kind of pathology of modeling. In the minds of modelers these networks grow in lucubratory complexity and in fascination, independently of their original intent to explain the brain … My own work has been focused on the olfactory system in small laboratory animals, because this is the simplest and phylogenetically oldest sensory system … Of the roughly 35 years that I have devoted to this study, the first 30 were spent in learning what questions to ask. Only in the past few years have some answers been forthcoming" (Freeman, 1991b).

He wanted to decipher the neural code using EEG. He argued that "perception is 95% expectation and 5% sensory input." His hypothesis was that information that exists in the system as an expectation is "a pattern of neural activity, on the basis of which the animal made the discrimination, and that this information would be detectable in some as yet to be determined properties of the EEGs" (Ibid). It remains to understand which pattern reflects which signal, and this will lead to cracking the code. Everything is logical. The question is in the tool. What did experiments give in 35 years?

"Our results showed that the information we sought was indeed manifested in the EEGs. It was identified as a spatial pattern of oscillation in the high frequency range of 40-80 Hz that we termed the "gamma" range, in analogy to the high end of the x-ray spectrum. A common wave form was found in each segment, and the spatial pattern of its amplitude tended to converge toward a reproducible shape each time that the background or an odorant was present. In principle this form of information is quite simple. It is like a frame in a black-and-white movie, in which the carrier wave is the light, and the shape is formed by the highs and lows of the amplitude or intensity of the light" (Ibid). Here's what the results of one of the experiments looked like:

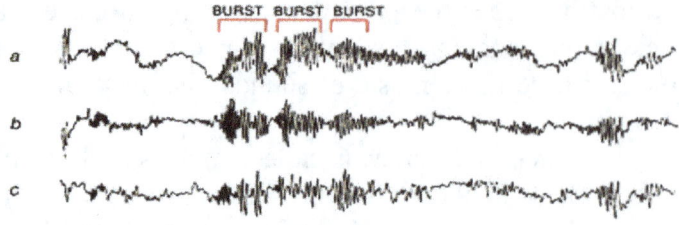

With such an analysis of the graph in the form of one melodic line, it turns out that bursts of high-frequency activity are observed when perceiving certain scents for which the animal has been trained. These bursts last a fraction of a second between the animal's inhalation and exhalation. Freeman found that the pattern repeats when the same scent is perceived. It looked as if the odor was encoded in the change in the amplitude of the EEG signal. It seemed that the neural code had been detected. Freeman wrote: "The shape of the carrier wave does not indicate the identity of the scent. That information is contained in the spatial pattern of amplitude across the cortex, which can be displayed as a contour plot, much like the plots of elevations in topographic maps" (Freeman 1991a).

Thus, using the equipment that, as he admitted, "contains little or no information in itself," Freeman assumed that the neural code is amplitude modulation of a specific high-frequency range. Based on this hypothesis, he analyzed phase portraits of the development of the EEG amplitude with different scents:

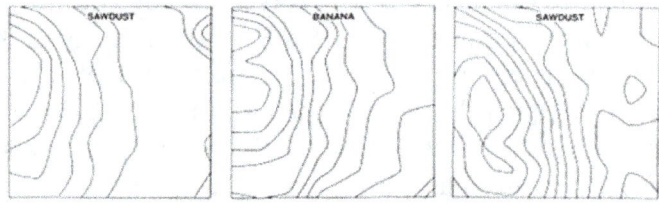

The lines represent the connection of points, each one being a set of three amplitudes of the EEG signal measured with an interval of a thousandth of a second. This topographic image shows how the EEG signal from the olfactory bulb of a rabbit forms a structure in response to the smell of sawdust, banana, and again sawdust. We can see that the sawdust pattern in the first graph differs from the structure in the last one, although we are talking about the same odor.

How did Freeman explain it? "Bulbar activity is dominated more by experience than by stimuli; otherwise, sawdust would always give rise to the same plot" (Ibid). It is not entirely clear: if we look at the EEG graph and believe that the amplitude modulation pattern contains information about the signal, then the same signal must give the same pattern if it is a neural code.

Yes, experience (projection) determines the perception of a stimulus (introjection). But representation cannot change all the time, and experience is all about repeating patterns. How then does the brain understand itself (reads its own code) and the world around if it has different patterns for the same signal all the time? And why does experience suddenly turn out to be a factor of instability and a complete absence of repetition, when intuitively, everything should be the other way around?

All this sounds strange, but there is an internal logic to such a paradoxical conclusion: Freeman is trying to interpret the results of his research, believing that amplitude patterns are the brain's code. The initial hypothesis generates numerous turns of thought and logical chains, but since the premise itself is erroneous, the conclusions become contradictory.

The question inevitably arises: are there any patterns in the system as representations of signals? Freeman did not find them and decided that they did not exist. The very word "representation" becomes an irritant for him. In the article "Representations: who needs them?" Freeman and his colleague proposed to forget about the term whatsoever for two reasons. "One is that no one now understands how brains work, but the use of the term representation and its attached concepts tends to obscure this fact. The term gives us the illusion that we understand something that we do not. We suggest that the idea of representation is seductive and enervating, promising good deals but delivering nothing new. When researchers refrain from using the term, knowledge of brain function is not significantly affected. We conclude that use of the term is unnecessary to describe brain dynamics. The second reason is that the use of the metaphor points us in a direction that carries physiological research away from more profitable lines of inquiry. We have found that thinking of brain function in terms of representation seriously impedes progress toward genuine understanding" (Freeman, Skarda, 1990).

The concept, which seems so intuitive (after all, why should the picture of the world not consist of representations?), is denied with such a passion of preaching: do not dare think in this direction; it is devilish temptation, seduction and corruption. What is the problem? Freeman did not find a mechanism for creating representations. He simply did not see any patterns in the EEG data. But who said that EEG patterns are the sought representations?

As is often the case, the lack of an answer led to the denial of the question. Freeman wrote about representations: "Who needs them? Functionalist philosophers, computer scientists, and cognitive psychologists need them, often desperately, but physiologists do not, and those who wish to find and use biological brain algorithms should also avoid them. They are unnecessary for describing and understanding brain dynamics. They mislead by contributing the illusion that they add anything significant to our understanding of the brain … In a word, representations are better left outside the laboratory when physiologists attempt to study the brain … Representation is like a dose of lithium chloride; it tastes good going down but it doesn't digest very well" (Ibid).

That is precisely the problem: physiologists have been trying to avoid functional, physical and technological questions. Looking for some special biological algorithms and mechanisms has led to an impasse that lasted for a century. The problem is not in the concept of representations, because that is what the brain is doing — representing the world, but in the "digestion" of physiologists, who cannot accommodate aspects required for understanding the brain dynamics: function, physics and technology. It is easier for them to throw the question outside the laboratory then to find the answer.

What did Freeman offer in return? "When many interconnected neurons within a mass of neurons fire together in pairs over repeated stimuli, the selectively co-active neurons are joined together into a network of strengthened connections. This conjoined set is called a nerve cell assembly (NCA), and we believe it is the basis for perception and learning in the nervous system. When perception takes place it is expressed in a reproducible and identifiable pattern of activity that is mediated by the NCA. When a novel stimulus is presented under reinforcement, a new NCA forms during the first few presentations. However, the background activity state of neural activity that goes on in the absence of the new reinforced stimulus should not be patterned, because this would drive the system into an already existing pattern of activity associated with another stimulus. The kind of activity that is required in order that a new pattern can emerge must be unpatterned yet controllable. This is chaos. It is generated by the nervous system in the presence of a novel stimulus so that the neurons have activity by which the Hebb rule can operate. Thereby a novel pattern can emerge from the chaotic state, and the system is not forced into preexisting "grooves" or patterns of activity" (Freeman, Skarda, 1991).

So, the brain activity aims to create reproducible and identifiable patterns, but this activity has to be unpatterned. What is the logic behind this self-contradictory statement? Freeman is perplexed by the question: if patterns do not disappear, how can new ones be created? It is a physical and technological question.

Several years before writing the article cited above, Freeman argued in another publication: "The diversity of forms and their experimental bases reflect the recurrence of the need for the concept of central nervous system representations and the breadth of the range of experimental data that support it. These inferences pose for the physiologist the questions, "What physical properties do representations have?" and "How can we explain the mechanisms of their formation?"… If by assumption a main function of the brain is to construct representations, then the key studies must be done as those representations are made" (Freeman, 1981).

Indeed, these are the very questions that the concept of representations poses to physiologists, and not only to them. But if the answer to these questions is not found, what is the suggested way out? Forget about questions and declare representations non-existent, and denounce the concept itself. Everyone has the right to change their minds, but what about the breadth of supporting experimental data? Should it also be forgotten and thrown out of the lab? We can remove the "non-digestible" concept but we have to offer something instead. And no matter what we propose, the same basic questions about physical characteristics and mechanisms will remain.

What is Freeman's answer? "We have found that an essential condition for these patterns to appear is the prior existence of unpatterned energy distributions which appear to be noise, but which in reality are chaos. New forms of order require that old forms of order collapse back into this chaotic state before they can appear. Therefore, in the EEG we see each burst appearing from chaotic basal activity and collapsing back into chaos, thereby clearing the way for the next burst of patterned activity" (Freeman, Skarda, 1990).

Well, let everything fall into chaos and rise from it like a phoenix from the ashes. But what is the physical mechanism? Without an answer to this question, the concept of "chaos" becomes a meta-metaphor and an auxiliary variable. The authors cannot get around this question. Authors' response: "The same neural system that generates bursts (signals) also generates the background state of chaotic activity (often thought to be noise). When the system switches (bifurcates) from chaos to burst activity, the chaotic activity stops and the signal starts. Chaos operates up to the moment of bifurcation" (Ibid).

These are terms borrowed from the dynamic systems theory. They describe changes in the phase portrait of the system, give a phenomenological description, but say nothing about the fundamental physical mechanisms. We can use them as a description, but we cannot pretend that it is an explanation. But for Freeman, this sounds like a self-sufficient explanation, and further physical details remain "behind the scenes" of his concept. Chaos becomes an analogue of God, which creates everything and destroys everything in a god-only-known way, or rather through the bifurcation of chaos into order and vice versa.

But where does the whole story about the emergence of patterns out of chaos and disappearance into chaos come from? The observations of EEG signals in the olfactory bulb, which is the primary filter-converter. The analysis of EEG data made by Freeman and colleagues based on amplitude patterns may have nothing to do with actual representations. And the zone itself may not be a place for creating representations, but only the first step of a hybrid analog-digital chain. But based on data from this zone and their dubious interpretation, Freeman concluded that there are no patterns in the brain as representations of environmental signals.

Okay, let it be chaos. Still, how does the system understand itself if the patterns are changing all the time? The same sawdust is different for the poor rabbit all the time. How can such a rabbit distinguish it from a banana? But it distinguishes. This is evidenced by our everyday life experience and by Freeman's experiments: the olfactory bulb in the rabbit's brain produces coherent patterns in response to a signal. The mystery is that these patterns can be different, but the signal is the same.

The answer to this puzzle is in the musical analogy and physics of the process. The same music can be played with different amplitude patterns. And no matter how much we study these patterns, even using complex methods of analyzing the phase portrait, we cannot understand music because completely different parameters determine it. And then we have to rely on chaos, as on some kind of omnipotent supra-entity that somehow creates this music.

But chaos is not some special mechanism, but the dynamics of the states of a complex system and its phase portrait in the form of a strange attractor. Freeman uses this term in this sense in contrast to unpatterned white noise (chaos in the ordinary meaning) and a simple phase portrait of a periodic signal in the form of a circle or a spiral. It is a way of describing the states of the system but not a description of the mechanism leading to these states.

Like the word "self-organization" does not mean an explanation of the mechanism but only carries the meaning that the observed phenomenon has internal laws that lead to manifestations of structuring and organization in a stable system.

At the same time, it is impossible to disagree with Freeman's conclusions about the dynamic nature of the brain and the complexity of its dynamics of changing pools of attractors. But all this remains metaphors for a general description of the observed phenomenon without disclosing the very mechanism underlying the observed complexity.

Here is how Freeman described brain oscillations and their interaction: "Each has its own characteristic period, but they differ and cannot agree. Neither can escape the other, so that perennial aperiodic oscillation results ... We say that for each learned class of stimuli the model has a chaotic attractor with a basin of attraction ... The commonly used sobriquet "self-organizing" connotes the feature that chaotic systems can create information as well as destroy it ... In brief, perception and memory recall form a unitary dynamical process by which meaning is created" (Freeman, 1991b)

But what is the physical mechanism for information creation and self-organization? How do patterns interact if they cannot agree in any way? Here the physical meaning is completely lost behind the fog of words about chaos and system dynamics. How oscillators, that cannot agree, create anything, any structure, not to mention the Mind and its meanings. There is no answer in the model, except for the constant repetition of mantras about chaos and its miraculous power.

Why such a strange idea that everything is created from the impossibility of oscillators to agree? All this comes from reading neumes of musical notation called EEG and ignoring the physical meaning of synchronization as a frequency and phase coupling of oscillators. Freeman even states that phase-locking and phase coherence notions cannot be applied to continuous processes as "the phase of a continuous frequency distribution cannot be defined" (Freeman, Skarda 1990b).

We cannot fix a phase of a continuous process, but we can define it. Moreover, all interacting oscillations of this Universe define each other phase; otherwise, they could not interact. But for Freeman, oscillations that have various periods (frequencies) cannot agree in principle.

We recall that the EEG shows a picture that is not differentiated by the cross-frequency relations and by the rhythmic phase structure of multiple frequency levels. Thus, it becomes clear that "non-agreement" seen by the researcher is due to the fact that the tool does not provide any information about the frequency and phase coupling at the very moment of interaction when the oscillators at different frequencies and different phases actually agree. The researcher does not see all the polyphony and polyrhythm of the music and judges what is observed: the disappearance of a more or less periodic simple pattern during background (spontaneous) activity and the appearance of a complex pattern of evoked activity during stimulus processing. In this mixed picture, we can see changes in the wave amplitude.

What conclusion can we draw from such a neume notation? The amplitude pattern is where the meaning lies. And we will study its dynamics and strange attractors, and we are bound to find nothing except chaos, that is, non-periodic dynamics that we see from the start without even going into complex analysis. We can do it for 35 years or longer with the same result.

If we take any dynamic system and try to analyze one parameter in relation to the narrow spectrum of another parameter (this is what the EEG shows), then we will inevitably get a complex phase portrait in the form of a strange attractor and the dynamics of the portrait change. Let's take a musical analogy again. If we take a narrow frequency range even in a simple piece of music and build a spectrogram and a phase portrait of the change in spectral power in this range, there will also be a change in complex attractors, since the amplitude may change non-periodically. We analyze the amplitude change graph, showing either a repetitive or a changing pattern even in the same melodic, harmonic and rhythmic structure. The meaning of the music is in this structure, but we are deaf to it. To complete the confusion, the amplitude pattern can be the same with a different musical form. As in studies of the rabbit's olfactory system: the sawdust is the same, but the EEG readings have a pattern that repeats and then changes.

If we ignore music in which both frequency and amplitude parameters and phase trajectory change, we might think that changing amplitude produces meaning. And since this change has complex patterns, we can say that this is the chaos that creates order. But how does it do it? Here we find ourselves again at a dead-end, like many others who asked how order arises from chaos but ignored the laws of oscillatory processes and their interactions. It is an impasse for physicists, biophysicists and physiologists.

Freeman's weather analogy can also be used. He compared EEG readings with measurements of airflows, behind which are temperature, pressure, and other parameters. Suppose we register the air movement in a particular place (analogy of EEG measurements) and simplify the system to one parameter of the flow power. In that case, we can compose differential equations of this dynamics and draw attractors. But will it give us an understanding of the weather? After all, in addition to the power, it has many more parameters. And this power itself is determined by deep processes with their own parameters.

The analysis of this power in one place and a narrow speed spectrum will give a complex dynamic picture of chaos with changing basins of attractors and patterns, and the theory of weather can be formulated as self-organized chaotic activity of airflows. But will this be the answer to the question of how does it work? No.

Similarly, there is no answer to the question of how does the Mind work in such statements: "We suggest that the self-organized process that replaces environmental input with an internally generated, chaotic activity pattern is one that gives "biological meaning" to the stimulus" (Freeman, Skarda, 1990b).

Whether we use the words "representation of signals" or, in trying to avoid the poisonous term, say about "biological meaning of the stimulus," the question remains: how meanings are produced, reproduced and combined into a coherent model of reality physically and technologically. Is it possible that the entire mechanism is in the amplitude modulation (AM)? In the previous parts of the study, we have considered the hypothesis that the brain uses all options since to create a broad palette of music of the Mind, it has to apply pulse-frequency modulation (melodies and harmony), pulse-width and pulse-position modulation (duration and sequence of rhythmic structure). This is how the Symphonic Neural Code is created.

Let's consider a purely engineering and technological problem of coding a message using only amplitude modulation. It was the very first technology for transmitting messages by radio. The carrier wave changed in amplitude following the code of the transmitted signal, and the receiver decoded the amplitude pattern into other parameters and received a representation of the original signal. This has also been used in analog television to encode the intensity of light applied to a screen. But if we want our TV to be not only black and white and radio broadcasts to carry a wide range revealing the original signal in full, then this method of encoding and transmission is not enough.

If we want the resulting representation to be of high fidelity, then AM will not be the preferred choice. And if we develop our technologies for coding and signal transmission, it cannot be the only method. For a simple reason: it has a lot of flaws. First, it has limited possibilities of encoding a wide range of parameters. Second, when receiving such a signal, the receiver is forced to catch together with the message a large amount of noise and interference, directly proportional to the signal itself. To improve the signal-to-noise ratio by 10 decibels, we need to increase the transmitter power by ten times. Third, AM is energetically ineffective as about 2/3 of the energy is spent on the carrier wave.

Frequency modulation (FM) is radically more efficient. This led to an almost complete replacement of AM when transmitting complex signals. But there is more: in a hybrid chain, digital modulation is possible, as a combination of a DAC (conversion, modulation, coding) and an ADC (inverse conversion, demodulation, decoding), where the energy efficiency, bandwidth and accuracy of the original signal representation comes to a completely different level. But this is the essence of the brain's function (creating the most accurate representations of signals, an adequate model of reality) and the essence of the evolution of the Mind (improving the functional efficiency and increasing the range of the model).

Is our brain such a backward system that has not advanced further than "black and white TV" for billions of years of evolution? Has the brain remained at the level of the narrow and ineffective "AM radio channel"? Does it put so much energy into the high-frequency range for this carrier wave to encode a minimal set of signal parameters in its amplitude? No. It processes and encodes a vast range of environmental signals, and in terms of efficiency, speed, range, it will give a head start to any artificial technology. How is it that neuroscientists' theories contradict the engineering and biological logic of survival and adaptation? How can a living system not drown in the potential infinity of signals if it is left with a pitiful way of encoding and transmitting meanings in the form of amplitude modulation of the high-frequency range? Where does this engineering nonsense come from?

Freeman began his education as an engineering student at the Massachusetts Institute and worked as a radio operator during World War II. Then the AM radio signal prevailed. Perhaps this is where his ideas have their roots? But he developed his brain theory when engineering had gone a long way from old methods. Is it the shortcomings of the EEG technology itself? But he was well aware of them. But in one of the articles, in the section "A look to the future," he wrote that of all the technologies known at that time (EEG, MEG, microelectrodes, optical technologies, different variants of tomography with registration of metabolic processes), EEG is "the most useful, inexpensive and reliable" (Freeman, 1991b).

All technologies have advantages and disadvantages. But it is essential what questions the researcher poses when using the tool. Freeman wrote: "The aim should be kept clearly in mind of understanding brain dynamics, not finding new methods of diagnosis and treatment" (Ibid). But understanding the dynamics of the brain is needed for diagnosis and treatment. There simply cannot be any other aim. Otherwise, research would be idle curiosity. We strive to know the device to be able to fix or even enhance it.

Freeman wrote about the pathologies of the Mind in the following way: "These maladaptive behaviors have some of the stigmata of dynamical systems gone pathologically awry in the brain ... No organic pathology in a contemporary sense can be attached to these conditions, nor should that be expected, if they are disturbances in some basic processes of self-organization by the brain" (Ibid).

Indeed, many pathologies are not associated with organic disorders detected by modern methods. We will consider this issue in detail in the following parts of the study. Applying the analogy with artificial information processing systems, we can say that systemic disturbances in the work of the Mind are often associated not with the breakdown of hardware (circuit elements, connections) but with software failures. Without knowing the technology and algorithms of the brain, we try to detect problems in the form of apparent changes in neural populations and "wires" (this is how traditional neurology works). In the absence of such changes, we begin to pronounce "mantras" in the form of pseudo-diagnostics describing external symptoms and not determining the state of the target organ (this is how psychiatry works). As a result, none of the systemic pathologies of the Mind is currently curable.

But if our model of brain dynamics remains at the level of phrases about "the self–organized process with an internally generated, chaotic activity pattern," then our understanding of the pathologies of these dynamics will remain at the level of phrases about "disturbances in some basic processes of self-organization by the brain." To avoid such general formulations about everything and nothing in particular, we need to know what basic processes work and how they normally work to understand what kind of violation can be associated with maladaptive states.

Let's try to look at the results of Freeman's experiments carried out on the olfactory system for decades with the aim of understanding the general patterns of brain dynamics. This is how the author summarizes his observations:

"There are five aspects that are very surprising. First, the information is uniformly distributed over the entire olfactory bulb for every odorant. By inference every neuron participates in every discrimination, even if and perhaps especially if it does not fire ... Furthermore, we and many other investigators have attempted to demonstrate odorant specificity in the discharge rates of action potentials from single neurons. These attempts have never succeeded ...

Second, we find that the bulbar information does not relate to the stimulus directly but instead to the meaning of the stimulus ... The regularities that we should seek and find in patterns of central neural activity have no immediate or direct relations to the patterns of sensory stimuli that induce the cortical activity but instead to the perceptions and goals of the subjects.

Third, the carrier wave is aperiodic. That is, it does not show oscillations at single frequencies, but instead has wave forms that are erratic ... Yet the same wave form appears on all the channels, although of course at different amplitudes. Moreover, during exhalation and prior to each inhalation there is persisting background activity ... Odorant segments in many instances show no increase or even a decrease in overall EEG amplitude from this basal state, yet always display a remarkable commonality of wave form across channels ... Indeed, a substantial aperiodic basal activity persists after the bulb and olfactory cortex are surgically de-afferented and isolated as a unit from the rest of the brain ...

Many parts of the physicochemical brain are capable of generating controlled but locally unpredictable activity that looks like noise but is not ... Our EEGs are showing us sequences of patterns, each of which is carried by a chaotic wave form and not by a wave at a single frequency, as previous EEG studies had led us to expect.

Fourth, we find that a surgical, pharmacological, or cryogenic inactivation of the main pathway from the bulb to the olfactory cortex causes the cortical activity to go silent both in respect to action potentials and to EEGs, and the bulbar activity to display nearly periodic oscillation with each inhalation. On recovery from the interruption of the pathway, the normal action potentials and EEGs return. These findings demonstrate that the background unit and EEG activity between inhalations as well as the induced impulse and wave activity during inhalations are global properties of the entire olfactory system ... The background state is not a silent equilibrium that is perturbed by noise. It is a chaotic state that keeps the system in constant readiness to jump to any desired perceptual state at any time.

Fifth, we found that pattern formation can be chemically controlled ... The modulations are implemented by well known transmitters and neuromodulators" (Ibid).

Let's try to deal with the "surprising aspects," applying hypotheses about the brain dynamics in the framework of TTT. The first observation will be surprising if we ignore the importance of the rhythm as a pattern of "sounds and silence" with a certain sequence and duration. Neurons are silent as meaningfully as they speak. The code cannot be and is not just firing. And although such an idea of the code as firing rate is an informational, mathematical, technological absurdity, that, to top it all, also contradicts the observed phenomena of the system's operation, it stubbornly lives and thrives in the models and experiments of many neuroscientists. Hence Freeman's surprise: no matter how hard they tried, they could not find a meaningful connection between the external signal and the neuron firing rate. They were looking for a black cat in a dark room that had never been there. But a negative result is just as important as a positive one. Perhaps it will move our thoughts in the right direction?

Now Freeman's second surprise. The carrier wave turned out to be aperiodic. This is a surprise if we proceed from the hypothesis that coding is by amplitude modulation. In that case, a carrier wave should be as periodic and stable as possible and strive for a sinusoid of an ideal oscillation. In AM, an encoded signal is superimposed on this sinusoid, and the transmitted meaning is contained in the change in amplitude of this sinusoid:

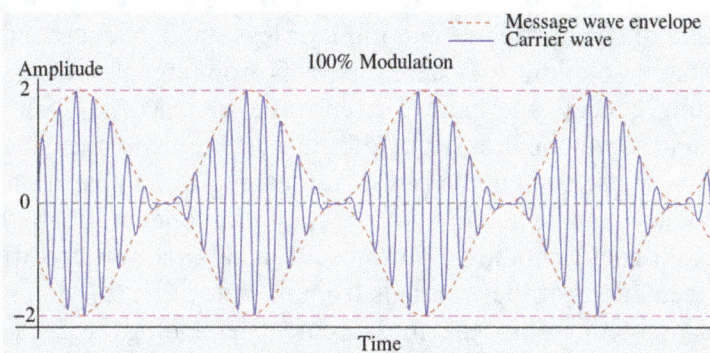

If the carrier wave is chaotic, no information will be transmitted. For Freeman, such an aperiodic carrier wave, that does not have one frequency, changes shape, does not repeat itself in time, can somehow carry information encoded in its amplitude. By the way, he notes the repetition of the pattern in space: on the channels (EEG sensors), the shape is preserved. This tells us that the wave carries information. It is no surprise at all. This is how it should be if we proceed from the hypothesis of the wave nature of the process.

But in what wave parameters is this information? If the amplitude characteristic changes unpredictably and is independent of the signal, if the wave cannot perform the carrier function, what should the conclusion be? That the hypothesis about the central role of amplitude modulation of the high-frequency range failed. Not to mention all the disadvantages of amplitude modulation itself, which makes it not the most desirable candidate for the role of a code initially.

But Freeman does not draw that conclusion. For him, this empirical evidence is not a negative confirmation (refutation) but a "surprising aspect." And even the fact that the odor segments, as a signal encoded by the system, do not show a change in amplitude compared to the basal state, when there was no smell, does not raise the question of the plausibility and correspondence of his model to empirical facts.

What is his conclusion? We have to look for a non-periodic carrier frequency and forget about looking for it at some specific frequency. But to make technological sense, the carrier frequency must be at a particular frequency. Maybe there is no carrier frequency for the amplitude modulation invented in Freeman's theory of a neural code? If the cat stubbornly does not want to be found, perhaps it is simply not there? But for Freeman the cat is there, but it begins to dissolve and take on chaotic form.

What about the basal state? Not only do baseline oscillations continue during rest, but they also occur in areas that are physically cut off from each other and isolated from the environment's signals. And they certainly are not noise. Freeman calls this preparing to "jump" from one pattern to another. But what kind of preparation and what is the mechanism of "jumping" remains outside the frame of the theory, except for the reliance on the magical power of chaos, which allows the system "to jump to any desired perceptual state at any time."

Now a little about the physics of the observed process and how this physics is reflected on the EEG screens. Researchers observe a mixed aperiodic picture with active coding of signal streams and more or less stable background oscillations in the basal state (isolation of system elements from signals). Does this mean that there is some general aperiodic carrier wave for AM? No. This suggests that researchers observe the transition of the system from the state of the basic synchronizing pulse (this is the physical meaning of "preparing to jump") to the state of a combination of numerous frequency components and rhythmic structures in the process of active coding of introjected signals and superposition of projected wave representations of these signals from models of reality.

But the problem is that the measurement technology used simply does not provide this complex picture with component differentiation. It mixes everything, which gives the impression of one, but aperiodic and unpredictable wave. And it is this heap that researchers are trying to make sense of. The problem is that this is a neume musical notation, in which it is impossible to reflect all the melodies, harmonies and rhythms of complex music. It gives only an overall picture of dynamics. It is useful, but not for trying to read unknown music. Here is such an unsurprising result that surprised Freeman.

The fourth surprise is also about the same basic pulse that persists despite all the efforts of researchers to deactivate the system. They did everything except killing a rabbit, but the system continued to oscillate. Freeman's conclusion is as logical as it is obvious: this is clearly a global characteristic of the entire system; it is present in both superficial and deep chains. But what is it? A kind of chaotic state that keeps the system in readiness for the jump? It's about everything and nothing again. How does it keep? How does it jump? It is clear that attractors and their basins replace each other. But behind these words, which describe only our representation of the oscillatory process in the form of a visual picture of a phase portrait, there are fundamental physical mechanisms of interaction of oscillations. There is not a word about this in Freeman's model.

The fifth surprise is not surprising at all, and here Freeman exaggerates a little. The fact that modulation of neuronal activity is carried out, among other things, by chemical compounds (transmitters and neuromodulators) is no secret for almost a hundred years: for the discovery in 1921 of the mechanism of the chemical transmission of nerve impulses, Otto Loewi received the Nobel Prize. But Freeman has an important clarification: it is not just about "transmission of nerve impulses," as the Nobel Committee formulated it many years ago, but about the formation of patterns. These substances are indeed active elements in the regulation of oscillatory activity to form and transmit information in the system. They help the parts of the system to "agree" at different frequencies and phases during the synchronization process. Then Freeman's fifth surprise becomes understandable: after all, in his model, the oscillators cannot agree in any way and fall into the pool of chaos.

As the authors of the review article "The brain as a dynamic physical system" noted, the central questions in describing the nervous system within the framework of the approaches of the theory of complex dynamic systems are: "Under what conditions and by what means does the organism use this attractor switching? …The ability of perturbations to switch dynamic modes in a phase-sensitive manner brings us to the general issue of how do complex systems move through multiple dynamic states?" (McKenna, McMullen, Shlesinger, 1994).

Indeed, if we are talking about the physics of a process in a physical system, then the main question remains: "How does it work?" Without an answer to it, any description method with the help of the most fashionable new terms or the old biblical ones remains metaphors either with an explicitly mystical bias as in religious terminology or implicitly mystical. For one simple reason: without a description of a physical mechanism that is realistic and testable, any description will be mysterious, even if it is full of physicalist terminology. The phrase "attractor switching" is fundamentally no different from "spirits moving." Both are suitable for describing the state of the system at a metaphorical level. For concrete answers, it is much more helpful to move from metaphors to physical analogies.

As the same authors wrote, many analogies have already been used when describing the system: from batteries, electrical networks and telephone switches to computers. But any analogy is applicable only if it helps to reveal the mechanism of operation of the described system. "The real issues are identifying sequences and transitions between dynamic states or attractors structures in brain systems, identifying what controls transitions between these states and determining the extent to which dynamic system analysis will permit us to find new candidate codes and representations for behavioral and perceptual sequences" (Ibid).

The question again and again comes to the neural code and representations, to the mechanism and algorithm of dynamic self-regulation of states. "How can single neurons which operate in the millisecond range reliably encode microsecond differences? …. Data suggest that the solution is found at the neural ensemble level, and involves the systematic phase shift of population oscillations

... The use of resettable and synchronizable oscillators may be a broadly used neural phenomenon ... The neural ensembles oscillations, per se, have been somewhat sceptically received, perhaps in part because reductionist neurophysiologists are comfortable with bursting in individual neurons than with waves in neural ensembles ... In fact, it has been suggested that the oscillations are epiphenomena ... Indeed, aperiodic activity with a wide spectrum is a common observation in many, if not most, neural regions. For those who wish to characterize the non-linear dynamics of neural activity, this observation is more challenging than periodic oscillations, but also more interesting ... The spatiotemporal pattern of activity in the brain is a fertile area for dynamic system analysis ... This new approach requires state-of-the-art recording techniques combined with the appropriate analytical tools for dynamic system characterization ... From the view of non-linear dynamic systems, neural subsystems are being coordinated or synchronized. Key factors ... are the phase shift and its systematic control, the intermittent nature of the synchronization and the identification of the waveform identification and its modulation across the spatial dimensions of the neural array" (Ibid).

In one sense or another, all neuroscience theories have been about patterns of brain activity. The question is which patterns are considered as candidate codes. Freeman criticized the fixation of many researchers on the activity of individual neurons and called this approach "pulse logic models" (Freeman, 1972). He unequivocally denied the need for such an analysis: "We are assured that the proper element for our model is the local population and not the single neuron. Perhaps paradoxically this provides an enormous simplification, because the dendrites of neurons in a local neighborhood can be modeled as a simple integrator ... Most of the complexities of the single neuron remain at the hierarchical level of the neuron" (Freeman, 1991b).

In his model, there is no place for the information carried by the activity of one neuron, and the order of spikes has the nature of a Poisson distribution (a random value of events independent of each other that occur at a particular time with a certain average intensity). Poisson distribution has the following features: it refers to events that do not have a causal relationship; the average flow rate should be constant, without accumulations, peaks and troughs; two events should not happen simultaneously; the probability of an event at a specific time interval is proportional to the length of the interval.

Do all these signs relate to the work of the nervous system? No. The work of neurons has a mutual cause-and-effect relationship, otherwise, there simply would be no consciousness as a result of network activity. The flow rate is variable. A huge number of simultaneous events take place in a neural network. There is no direct proportional relationship between the probability of neuron activity and the time interval. And the most important thing: the activity of neurons has a structure, not a random distribution. If EEG readings do not show this structure, it does not mean that it is not there.

In Freeman's model, neurons are "sparsely drawn on at random to fire, and convert neural current strength to pulse probability" (Freeman, 1992). That is, the firing of neurons is discrete pulses without any information density of the parameters of the pulse itself, and all these pulses randomly create a wave structure in which there is a pattern of information.

"The probability of firing of each neuron is determined by the local value of this wave, superimposed on background activity, but the same wave occurs in many neurons in the same domain. The sum of probabilities of firing over those neurons gives rise to an actual pulse density (number of pulses per unit area per unit time), which is the output ... In terms of its accessibility to observation with a microelectrode, it is a pulse function — that is, a set of near-random pulse trains with some coherence in terms of mean firing rate in relation to an afferent stimulus" (Freeman, 1972).

This is how the physical meaning of the process is described. Waves form by the probability of random discrete shots joining them. Indiscriminate firing creates a wave of meaning, and the only sense of the firing itself is its average speed, which encodes the incoming stimulus.

This strange picture resembles something. A description of the physics of the process is reduced to a statistical description of the probabilistic process of creating forms and structures from a sufficiently large number of randomly interacting discrete elements. It is the main slogan of statistical physics of the twentieth century: order from chaos through quantity. But as we remember, it did not lead to answering the question of how this order arises.

It is pertinent to recall here that Schrödinger, when trying to transfer the idea to biological systems, reasonably noted that there are simply not enough elements for the statistical principle of the emergence of order from chaos to work (Schrödinger, 1944). Schrödinger himself resisted the probabilistic interpretation of his wave formula for a very long time but eventually gave up, and physical waves became waves of probability.

The brain waves in Freeman's model are also described as such. It is, in essence, the Copenhagen interpretation of the brain wave function. Apparently, for him, the number of brain elements was enough to form order out of chaos. In physics, this approach was a reflection of the helplessness of concepts when it came to describing the physical mechanism of creating energy-matter structures and their interaction. If there is no mechanism with regularities, then it remains to randomly "roll the dice" of probability, and everything will be formed by itself sooner or later with varying degrees of probability.

The problem is not in statistical and probabilistic methods. By themselves, they are the only ways to describe a large number of signals. The problem is that statistical, mathematical tricks replace the physical phenomenon and its mechanism, and probabilistic statistics becomes not a way of describing but the creator of matter and the Universe.

Is it a coincidence that we see the same tricks in explaining the work of the brain? Of course not. After all, we are talking about matter and the material process in it. We discussed the microcosm of basic energy levels and the macrocosm of universal scales, and here we are talking about the intracellular, cellular and intercellular level of biophysics. The levels are different, but the problems in the description are the same. Perhaps the ways out of these dead-ends will also be similar? A musical analogy will help.

Here's Freeman's description: there is a large set of randomly sounding identical notes; from this chaos, probabilistic combinations arise, which become the meaning of music, its melodies, harmonies and rhythms; identical notes have one way to create these structures — to fire randomly and independently of each other (Poisson distribution), so as to convert the current strength and average speed into meaning patterns; all this will lead to waves of probability that there will be music. How likely are we to create music this way? There is always a probability, it is never equal to zero, but with such a mechanism of creation, it tends to zero with a high average speed.

If we use a musical analogy, then the strangeness of this version of the neural code becomes even more apparent. But where does it come from? Again, we must remember the musical notation that Freeman looked at. The neumes were so approximate and mixed that the researcher's conclusion about randomness of pulses is not surprising. Freeman's model is both the result of a conceptual error and drawbacks in measurement technology. EEG may be inexpensive and reliable, but its usefulness in studying complex patterns of neural activity is questionable. More sophisticated methods of measuring using implanted electrodes and a detailed study of the temporal structure of the spikes and interspike intervals show that it does not have the Poisson type of distribution.

The authors of one such study wrote: "Our findings are inconsistent with Poisson models of spike trains ... Despite the great variability in the discharges of cortical neurons, the spike-generating mechanisms are intrinsically very precise. Such precision is necessary for the propagation of information by a high-resolution temporal code ... Our analysis of neural responses in macaque V1 and V2 demonstrates that each of the stimulus attributes studied (contrast, check size, orientation, spatial frequency, and texture) systematically change not only the absolute number of spikes, but also their temporal pattern "(Victor, Purpura, 1996).

Logic dictates, and facts prove, that random firing cannot create meaning. In the music of the Mind, both the pitch of the notes and the rhythmic pattern as the sequence and duration of notes and pauses are important. The number of notes, of course, also varies depending on the complexity of the piece, but in itself, it does not have sense. Even a huge number of neurons cannot create Mind if they are "sparsely drawn on at random to fire."

On the one hand, wave models like Freeman's theory talk about continuous oscillatory processes and are the opposite of models that believe that the neural code is created by identical discrete spikes. But, on the other hand, they also ignore the frequency and temporal structure of neuronal activity. Thus, they lose physical and technological sense and contradict the physiological realities of the brain. Neurons are trying very hard to create a high-resolution and high-precision code, and the researchers say: we don't see the point in the intricacies of details, so they do not exist. As the authors of the review correctly noted, the study of spatiotemporal patterns of brain activity requires state-of-the-art recording techniques. Present equipment does not yet allow us to compose a full-fledged musical notation of the music of the Mind. But our models should move engineering forward, not trail behind what inexpensive and reliable, but not useful enough equipment provides.

The problem of neuroscience is not technical but conceptual. What's the outcome? Let's use a quote from the "Neuroscience" textbook: "Clinical and brain imaging observations do not, however, provide much insight into how the nervous system represents cognitive information in nerve cells and their interconnections … The observations summarized here are merely a rudimentary guide to thinking about how complex cognitive information is represented and processed in the brain, as well as how the relevant brain areas and their constituent neurons contribute to such important but still ill-defined qualities as personality, intelligence, and other cognitive functions that define what it means to be human" (Purves et al., 2012).

It sounds like a statement of the complete failure of an entire branch of knowledge. Neither the study of the external manifestations of the brain work, nor the study of internal activity answered the most crucial question: what is the Mind, and how does it work?

Now let's recall that in this textbook, as in all the others, and in the field as a whole, there is no definition of the Mind from a functional, physical and technological point of view. How can we achieve an understanding of the object of research if we do not define the object itself? The object of cognitive science research is not the brain but the process taking place in it, the brain work that we call the Mind. Without defining this process, we are walking in a vicious circle. If something is poorly understood, then it does not give much insight into how the ill-defined works. If it is not defined, we have no understanding. If it is not clear, then nothing is definite.

There is only one way out of this circle: to take a chance and give a definition. It may be wrong or right, but we cannot judge if it does not exist at all. Let's get back to the definition on which the Teleological Transduction Theory is built:

The Mind is the process of transducing signals from the external environment and the body into the internal code patterns representing these signals and constituting a model of reality for the purpose of active adaptation to this reality and maintaining the integrity of a living system.

The process includes the following stages:

1. Encoding representations.
2. Integrating representations into a coherent model of reality.
3. Storing representations.
4. Assessing the state of the environment.
5. Managing external actions.
6. Assessing and regulating the state of the body.
7. Comparing projected model and current signals.
8. Creating and projecting an updated version of the reality model.

The physical processes listed in this definition constitute a system that we call Personality. Its coherence, integrity and adaptability depend on how these processes evolve. Here, the concept of personality is considered in the broad sense of the system's overall functionality and not just social aspects. We have already considered algorithms and technologies for creating representations and partially touched upon the issue of their integration into a subject's model of reality. In this part, we will continue to explore how a unified Personality is created from parts and stages.

Chapter 4

Personality as a Physical Process

From the brain, and from the brain alone, arise our pleasures, joys, laughter and jokes arise, as well as our sorrows, pains, griefs and tears.

Hippocrates

In the famous painting "The Creation of Adam," Michelangelo depicted a scene from the biblical text about the evolution of the world "Book of Genesis":

Usually, the plot is described as God sending a spark of life to Adam. But another interpretation, more suitable for the real scene in the picture and the biblical story, is possible. The Bible describes the creation of Adam as an evolutionary process: aquatic forms of life arise first, then plants, amphibians and reptiles emerge from the water, then birds and land animals appear, and only after them comes the turn of man. The Bible is not far from the modern evolution theory. It only skips protozoa and other simple life forms that were unknown at the time when the book was written.

The spark of life came from the ancestors on the evolutionary tree. The picture shows that Adam has already been created as a living being. He reaches out to receive another spark: human intellect that will allow Adam to "rule over the fish of the sea and the birds of the air, over the livestock, over all the earth, and over all the creatures that move along the ground" (Bible. Old Testament. Genesis. Chapter one).

A possible proof that Michelangelo meant the spark of consciousness may come as a surprise. One of the modern researchers noted that he depicted God inside a tissue that resembles a cross-section of the brain (Meshberger, 1990):

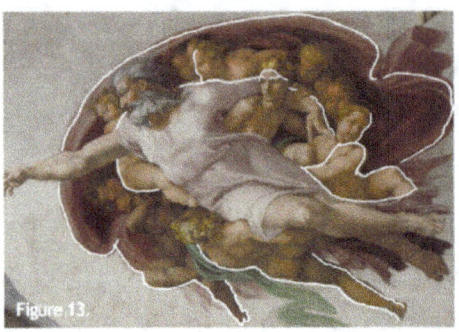

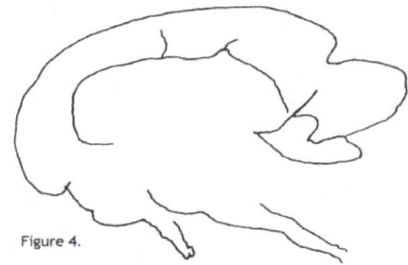

Simplified anatomical diagram of a human brain:

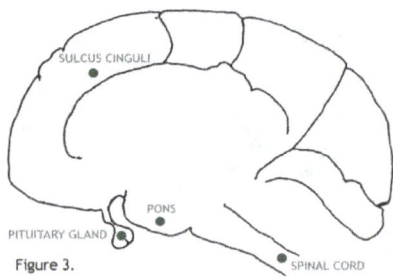

Meshberger, 1990

But such a version does not seem surprising if we take into account the historical reality. First, Michelangelo was not only an artist but a physiologist. To depict a human body, he had to know its external and internal structure. Second, whether the depiction resembling the brain is a coincidence or not, the fact remains that at the time of Michelangelo physiologists already knew well that the brain is the source of our thoughts, senses, pleasures and pains.

Many species in the animal kingdom have a nervous system. The further up the evolutionary tree, the more it is complex. The higher branches are distinguished by a structure called the neocortex. It is smooth in rodents and other small mammals, but higher mammals, including humans, have neocortex with deep grooves (sulci) and ridges (gyri). In humans, 90% of the cerebral cortex and 76% of the entire brain is neocortex (Noback et al., 2005). Here is a rough anatomical scheme of the human brain from a lateral view:

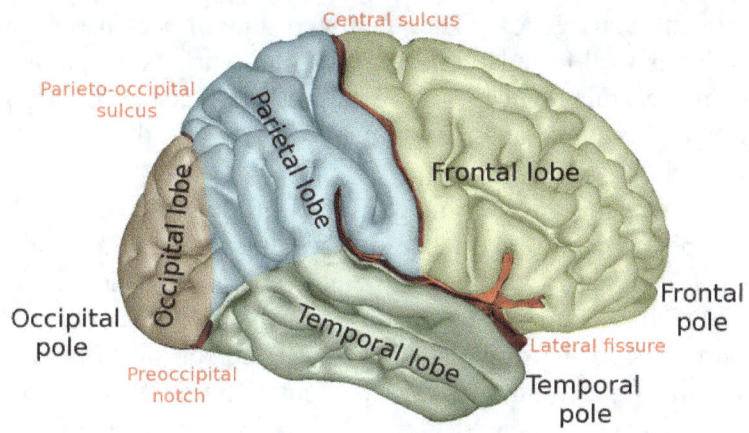

The normal human brain has areas that are absent or less developed even in our closest relatives on the evolutionary tree. These regions concentrate in the anterior part of the cortex. Within the technological brain map suggested in TTT, the cortex consists of filters-integrators. Moving from posterior to anterior regions we get from primary to intermediate and higher integrators that play a key role in functions that we typically attribute to human intellect.

Here we will not discuss the question of whether the emergence of these functions was a single act of God sending his spark or the result of a continuous process of development. For those interested in the first hypothesis which does not have any empirical confirmations, we recommend the Bible. For those interested in the second one, there are numerous empirically supported studies. A good review was done by Todd Feinberg and Jon Mallatt in an article called "The evolutionary and genetic origins of consciousness in the Cambrian Period over 500 million years ago."

Within the topic of this chapter, we are interested in the current state of affairs. The authors of the review describe it this way: "The creation of a sensory neural map requires at a minimum a brain and typical neurohierarchical structure, with consciousness emerging from progressively more complex and integrated patterns of isomorphic organization in the upper levels of this hierarchy. But the brain-maps for the different senses are not isolated from one another. They are integrated in two critical ways. First, the highest levels are multimodal … The second critical feature of the neural correlates of consciousness is widespread interaction … Hierarchically arranged neural-neural interactions created conscious sensory images and their associated qualia. This made sensory consciousness into the unique neurobiological system feature that it is" (Feinberg, Mallatt, 2013).

But there is one more important aspect of conscious awareness. As the authors noted, "the intricate, consciously perceived, sensory maps of an animal's outer world (and inner world) lead to improved information-gathering and better decisions on how to respond to complex and changing environments" (Ibid). The crucial addition to the model of external reality is the inner world model that we experience as a feeling of a unified I (Self, Personality). Personality has three aspects: an internal reference point for observation and action (bodily agency), a narrative of the observations and actions (reflecting agency), and a purposeful observer (intentional agency). The integrated state of each aspect and all of them taken together is vital for adaptation to the outer world. The major difference that made Adam the "ruler over all the creatures" is that reflecting and intentional agency are highly developed in humans. Where is this "spark" of intellect generated?

One of the ways to answer this question is to look at the correlation between the internal phenomenal experience of various states of Self and the functional states of various cortex regions. In one experiment, the researchers invited experienced meditators who could induce themselves into a decreased or increased feeling, observing, or narrative state. The first-person phenomenological reports and standardized questionnaires which focused on subjective contents of three aspects of Self were compared with the functional connectivity analysis of the EEG data.

The authors reported: "The main positive finding of this study was that every time participants mentally and in a controlled manner willfully up-regulated the expression of "Self" (witnessing agency), "Me" (body representational-emotional agency) or "I" (reflective/narrative agency), the functional integrity (indexed by EEG operational synchrony) of the corresponding self-referential network operational modules (SRN OMs) increased significantly, and conversely, wilful down-regulation of the "Self", "Me" or "I" expression resulted in a significant decrease of the functional integrity of the respective SRN OMs. The observed increases or decreases in the functional integrity of the SRN OMs were predictably coupled with participants' self-description of alterations in the phenomenological experience during up- or down-regulation of "Self", "Me", and "I" states, and significantly correlated with phenomenological factors estimated by a set of standardised questionnaires" (Fingelkurts et al., 2020).

To investigate these findings further, the authors studied the functional integrity of the OMs in a patient diagnosed with a condition called Depersonalization Disorder (DD). Phenomenologically this pathology is described by psychiatric manuals as the subjective experience of detachment. Patients feel like outside observers of their bodily sensations and thoughts. These symptoms are persistent and lead to an overall feeling of being an automaton.

There are some paradoxes in the condition. First, patients feel detached from the body but acknowledge it as their own on the rational level of thought. Second, the feeling of disembodiment goes along a normally functioning interoception. Third, the feeling of detachment from the world does not come from some disruptions of exteroception. Fourth, the feeling of emotional numbness is not the outcome of the disruption of basic emotional systems. Moreover, such people experience chronic anxiety and fear, that is, emotional activity is actually increased. Can these paradoxes be explained on the basis of the observed functional state of the brain?

The normal integrated state of Self is the product of the functional coherence of a complex network the cortex plays a major role. The authors define the cortical structures as operational modules (OMs) that include the anterior OM and symmetrical (left and right) occipito-parieto-temporal OMs. They propose that the frontal module is linked with the phenomenal first-person perspective (Self-module) when one feels at the center of perceptual reality and the sense of agency as ownership of thoughts, perceptions, and actions. The right posterior module is associated with the experience of self as localized within bodily boundaries entity (Me-module). The left posterior module is involved in the experience of thinking about and reflecting upon oneself, including inner speech and interpretation of episodic and semantic memory events (I-module). They measured the EEG functional connectivity of these OMs in a patient diagnosed with DD.

Here is the report: "We observed that subject with DD exhibited a profound reorganisation in the integrity of three SRN OMs (indexed by the EEG operational synchrony analysis). Such reorganisation was expressed through a strong enhancement of functional integrity of the Self-module of the brain SRN, moderate decrease of the functional integrity of Me-module and a considerable decrease of the functional integrity of I-module of the brain SRN" (Fingelkurts AnA, Fingelkurts AlA., 2022).

The authors call the increased integrity of the anterior zones the "hyper-observation" and explain it as a compensatory mechanism for the dysfunction of other modules. From the TTT perspective, the explanation looks somewhat different. The increased functional integrity of the frontal zones does not mean that they work better than in the normal state and compensate for something. There is a desynchronization and disintegration of the entire system responsible for the formation of an integral personality. Thus, the paradoxes of depersonalization are resolved. Detachment from one's own body and from the world, while maintaining a critical assessment of the state at a rational level, means that the structures of higher integrators work in isolation from primary and intermediate integrators. The integrity of the PAAL algorithm is violated. The projection of the general model of reality, including the model of the Self, carried out by the anterior zones does not intersect with the introjection of representations of current signals created by the posterior zones of the cortex.

This is the technological reason for the dissociative state, which can be natural, as in the case of trance (meditation), or pathologically chronic, as in the case of DD. We will consider such conditions and their physical causes in detail in the following volumes of the study. Here we only emphasize that in order to explain how the physiological system produces mental phenomena, it is not enough to label the structures of the substrate as operational modules that perform a particular function. The authors only give new names for the standard functional-anatomical map with its classification of the areas of the cortex as sensory-motor and speech structures of the occipital-parietal-temporal lobes and higher associative structures of the frontal lobes. We need to show the physical mechanism and the technological implementation of this mechanism, which is behind mental and physiological processes.

The cortex anatomy has been studied in detail since the end of the 19th century. Santiago Ramón y Cajal, one of the founders of modern neuroscience, drew a section of the cerebral cortex in this way:

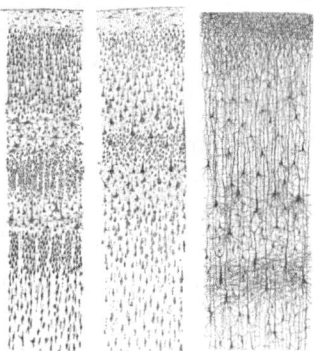

Cajal, 1911

The neurons of the cerebral cortex are very diverse in shape. They are called accordingly: pyramidal, stellate, spindle-shaped, arachnid, horizontal. In nature, nothing happens for no reason: a difference in form, as a rule, means a difference in function. Pyramidal neurons constitute the main and most specific form for the cerebral cortex (80-90% of all neurons). Dendrites extend from the top and lateral surfaces of their bodies, ending in various layers of gray matter. Axons go from the base of the pyramidal cells. In some cells, they are short, forming branches within a given section of the cortex; in others — long, entering the white matter (connections with other areas of the brain). Axons of large pyramidal neurons take part in forming pyramidal pathways that project impulses into subcortical structures.

The neurons of the cortex are arranged in layers with fuzzy boundaries. Each layer is characterized by the predominance of one type of cell. There are six main layers:

LI: The molecular layer of the cortex contains a small number of tiny fusiform cells. Their axons run parallel to the brain's surface as part of the tangential plexus of nerve fibers in the molecular layer. The bulk of the fibers of this plexus is represented by the branching of the dendrites of the neurons of the underlying layers.

LII: The external granular layer is formed by small neurons, which have a rounded, angular and pyramidal shape, and stellate neurons. The dendrites of these cells rise into the molecular layer. Axons either go into the white matter or, forming arcs, enter the tangential plexus of the fibers of the molecular layer.

LIII: The layer of pyramidal neurons is the widest compared to other layers of the cerebral cortex. From the top of the pyramidal cell, the main dendrite departs and locates in the molecular layer. Dendrites originating from the lateral surfaces of the pyramid and its base are short and form synapses with adjacent cells of the same layer. The axon of the pyramidal cell always departs from its base. In small cells, it remains within the cortex. The axon, which belongs to a large pyramid, usually forms a myelin fiber that goes into the white matter.

LIV: The internal granular layer in some cortex areas is highly developed (for example, in the visual zone). In other areas, it may be absent (for example, precentral gyrus). Small stellate neurons form this layer. It contains a large number of horizontal fibers.

LV: The internal pyramidal layer is formed by large pyramidal cells (Betz cells). Their axons form the central part of the corticospinal tracts and end on the motor neurons of the brainstem and spinal cord. Axons from giant Betz cells give off collaterals that send impulses to the cortex itself. Collaterals of the pyramidal tract fibers also go to the striatum, the red nucleus, the reticular formation, the nuclei of the pons and the lower olives, which transmit signals to the cerebellum. Thus, when the pyramidal tract sends a signal that causes targeted movement to the spinal cord, the signals are simultaneously received by the basal ganglia, stem and cerebellum.

LVI: Various, mainly spindle-like neurons form a layer of polymorphic cells. The outer zone of this layer contains larger cells. The neurons of the inner zone are smaller and lie at a great distance from each other. The axons of the cells of the polymorphic layer go into the white matter as part of the efferent pathways of the brain. Dendrites reach the molecular layer of the cortex.

Among the nerve fibers of the cerebral hemispheres, it is possible to distinguish: associative fibers connecting separate areas of the cortex of one hemisphere, commissural fibers connecting the cortex of two hemispheres, projection fibers connecting the cortex with the zones of the lower parts of the central nervous system, afferent fibers that terminate in the layer of pyramidal neurons.

This is how a standard functional-anatomic diagram looks like in a modern textbook:

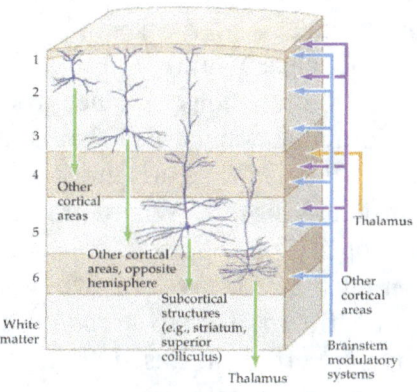

"Neuroscience," Purves et al., 2012

This diagram in a simplified form shows the outgoing and incoming connections of the cortical layers with subcortical structures and the connections of the layers with each other. If deciphered, we get loops of forward and feedback links. The outer loop connects the uppermost LI with the thalamus and the brainstem, which provides the links of the cortex with the spinal cord. Granular LII forms bonds of the upper molecular layer with the lower ones. Pyramidal LIII mainly specializes in connections between zones of the cortex itself. The internal granular LIV is connected with the subcortical structures, especially with the thalamus on the way to peripheral sensory systems. LV neurons are pyramidal with a large body and long axons extending both vertically and horizontally. They extend up to the LI and down to the thalamus, connecting LV with the midbrain, basal ganglia, cerebellum, brainstem and spinal cord. The axons of the LV layer are connected not only with those nuclei of the thalamus from where the cortex zone receives ascending introjection signals but spread to other thalamus nuclei. The axons of the lower, polymorphic LVI layer also go to the thalamus.

We get a clear picture not only of the central location of the thalamus but also of its central role. György Buzsáki, in his book "Rhythms of the Brain," dedicated a special chapter to the thalamus and called it "Thalamus: A Partner for Neocortex" (Buzsáki, 2006). Here is what he wrote about the concepts about the role of this critical area of the brain: "According to textbook wisdom, the thalamus is a large collection of relay nuclei, a kind of customs and border patrol agency. These nuclei are the only source of information for the neocortex about the body and the surrounding physical world. With the exception of olfaction, all sensory modalities are scrutinized by the thalamus before they can proceed to the neocortex. How the incoming stimuli are evaluated by the thalamus is quite a mystery, mostly because at first glance there is not too much coordination among the nuclei patrolling the different modalities. Even neighboring neurons cannot chat with each other directly, since they do not possess local axon collaterals, or only very sparse ones in some nuclei … The cytoarchitectural organization of the thalamus is unique. Unlike in the neo-cortex, where inhibitory cells are nested within the excitatory networks and adjacent to their targets, most GABAergic interneurons in the thalamus reside in a thin shell surrounding the thalamic chamber, called the reticular nucleus, and some other subcortical nuclei. The evolutionary cause or advantage of the spatial segregation of inhibitory and excitatory populations is not clear" (Ibid).

Indeed, if we approach the functions of zones from the point of view of only spatial location, then a lot remains unclear. What is the thalamus doing between the sensors and the cortex? It is logical to assume that it distributes flows. But why do we need a special structure for this? Why don't the streams just go directly to the specialized sensory zones of the cortex?

Energy consumption must be adaptive and evolutionarily justified. This is not a hypothesis but a fact of life: ineffective structures simply do not survive, and, therefore, those who live are efficient. Based on this common-sense logic, it can be assumed that there is some critical mission for the thalamus and other similar nuclei. But it remains a mystery, and, as Buzsáki noted, the spatial version of the function predominates: if it is in the middle, then it's a relay. This, of course, explains the multinuclear spatial organization. But why such complexity of the temporal organization, which is created by fine-tuning the dynamic parameters of neuron activity? Is it really just the distribution of flows in time? If the relay station has different paths, then it is not necessary to create a schedule to avoid traffic jams. All "trains" will simply go their own route and at their own pace. The fixation of researchers on the anatomical structure led to questions that had no answers. It seemed that nature complicated unnecessarily. Gradually, the understanding came that the dynamics of the network are needed no less than its topology. However, the emphasis still remains on the role of the thalamus as a connecting and distributing hub.

Buzsáki described: "With the thalamus as a matchmaker, the effective connectivity between local neocortical populations can be changed according to current computational needs. The key ingredient in this globalization process is the ability of the oscillatory mechanisms to recruit anatomically distant cortical neurons into temporal coalitions … Viewed from this new anatomical-physiological perspective, the thalamus is no longer a gigantic array of independent relays but a large communication hub that assists in linking large cortical areas in a flexible manner. The principal mechanism of the cortical-thalamic-cortical flow of activity is self-sustained oscillations" (Ibid).

But the question arises: do all computations take place only in the cortex, and the thalamus simply ensures the efficiency of the connection? It turns out that streams are distributed and connected, but what these streams are and how they turn into information remains a mystery. The absence of a technological approach creates an explanatory gap. Let's try to formulate a hypothesis that will proceed from the previous TTT hypotheses about the physics and technology of the process that we call the Mind.

Hypothesis:
The cortico-thalamic system (CTS) works according to the iterative principle of feedforward-feedback PAAL algorithm:

Physiologically, it is the connection of two chains: corticothalamic (CT) and thalamocortical (TC). The projection cortico-thalamic circuit involves the layer of pyramidal neurons LIII, the internal granular layer LIV. It then goes from the

internal pyramidal layer LV and polymorphic LVI to the thalamus nuclei, subcortical and peripheral structures. The introjection thalamocortical circuit extends from the thalamus to the LI molecular layer and the LII external granular layer.

Technologically, CTS is a modulating and integrating circuit. In the thalamus, the signal modulation occurs for the transformation of streams that have undergone primary coding in the filters-converters of sensory modalities into an internal format suitable for the filters-integrators of the cortex. Specific nuclei of the thalamus are responsible for the distribution and modulation of introjection streams. Non-specific nuclei intake the projection of integrated representations and command signals. The cortex layers and zones as primary, intermediate and higher integrators are responsible for receiving, distributing and binding introjection flows, forming separate representations and projecting the integrated model of reality.

Physically, CTS is a system of interacting oscillators that create unity and smoothness of cognitive processes, using the mechanism of frequency-phase coupling in the entire frequency range of the brain. This synchronization binds information flows into a unified model of reality and an integral model of I (Self, Personality). It is the physical basis of the state of awareness, which is regulated by the interaction between thalamic nuclei, modality-specific areas and higher integrators of the cortex. In case of disruption in this interaction (sleep, trance, pharmacologically induced dissociation, pathological conditions of the substrate) consciousness awareness fades into an unconscious state that has transitional levels depending on the scope of desynch.

The hypothesis not only shows the movement of flows in the network, but also explains them from a physical and technological point of view. We again see the manifestation of the universal PAAL algorithm, connecting the flows of the accumulated model of reality projection and introjection of current representations in the process of signal transduction and forming an updated model of reality.

Interestingly, the connections of the corticothalamic tract are quantitatively much superior to the thalamocortical and all other incoming connections to the thalamus from subcortical structures. For instance, the thalamic lateral geniculate nucleus (LGN) receives only 5-10% of the total inputs from the retina but gets a massive cortical projection from LVI. The pulvinar nuclei (PULV) receive only minimal afferents from the sensory periphery, and most of the inputs originate from the cortical LV and LVI. The projection part of the algorithm has functional priority and dominates even quantitatively (about 70-80% of all connections in the network).

The hypothesis includes both the connecting and computational role of the thalamus as a filter-modulator. It does not contradict the previous ones but eliminates the gaps left by them and, accordingly, is an explanation of a higher level. It not only answers the initial question of why would a system need a hub with modality-specific nuclei when there are modality-specific cortex zones. It removes confusion about the evolutionary cause and advantage of the high specialization of the thalamic network elements as excitatory and inhibitory populations. Activation-inhibition has two functions: regulation of the oscillation parameters and logic gates on/off state. Both functions are part of the signal encoding process.

Modulation is an intermediate technological step. The functional role of the thalamus coincides with its spatial location at the intersection of flows in the diencephalon (mid-brain). By the way, the olfactory system mentioned by Buzsáki as an exception also has intermediaries in the olfactory cortex, which consists of the anterior olfactory nucleus, the piriform cortex, the olfactory tubercule, and other structures. They perform modulation and primary integration functions before sending data to the frontal cortical areas for multisensory integration. The olfactory cortex is also connected with the thalamus.

The circuitry going through the thalamus is so complex that it may take a separate volume to describe it. Here we will just take two examples.

The lateral geniculate nucleus (LGN) is the key component of the visual modality. The excitatory input from the sensors of the retina is a small part of the circuit. The rest is modulatory local inhibitory circuits, descending inputs from LVI of the visual cortex and ascending inputs from the brainstem. "These modulatory inputs control many features of retinogeniculate transmission. One such feature is the response mode, burst or tonic, of relay cells, which relates to the attentional demands at the moment. This response mode depends on membrane potential, which is controlled effectively by the modulator inputs ... The thalamus sits at an indispensable position for the modulation of messages involved in corticocortical processing ... Thus, the full impact of the thalamus may be much more than simply controlling flow of information from the periphery and from other parts of the brain to the cortex: it may remain an active partner in all cortical computations" (Sherman, Guillery, 2002).

The reticular nucleus (RTN) occupies the central place physiologically and functionally. It forms a shell around the thalamus connecting the specific and nonspecific nuclei and the cortex. Nonspecific nuclei have converging input from the cortex, inhibitory inputs from RTN, and input of neuromodulators coming from several brainstem centers. Inhibitory signals from the RTN to specific thalamic nuclei balance excitatory signals from the cortex having a modulatory effect on the target cells. RTN is associated with the dorsal thalamus (DT), which receives signals from the periphery (external and internal sensors) and subcortical structures. The DT sends activation signals to the RTN which has inhibitory feedback to the DT, thus creating a chain from the cortex to the RTN and DT. All streams are either GABA (inhibitor), AMPA (activator), or NMDA (wide profile) regulated. The feedforward-feedback loops with activation-inhibition are the key to creating and maintaining synchronization.

"The capacity of relay neurons to operate in different voltage-dependent functional modes determines that the inputs from the cortex have the capacity directly to excite the relay cells, or indirectly to inhibit them via the RTN, serving to synchronize high- or low-frequency oscillatory activity respectively in the thalamocorticothalamic network ... Interactions of focused corticothalamic axons arising from layer VI cortical cells and diffuse corticothalamic axons arising from layer V cortical cells, with the specifically projecting core relay cells and diffusely projecting matrix cells of the dorsal thalamus, form a substrate for synchronization of widespread populations of cortical and thalamic cells during high-frequency oscillations that underlie discrete conscious events" (Jones, 2002).

Any violation in the activation-inhibition mechanism in CTS leads to disruption of the cognitive process and various pathological states of consciousness. This has been confirmed by numerous studies and there is even a special term "thalamocortical dysrhythmia" (Llinas et al., 1999). Any pharmacological intervention into the oscillatory process also results in disruption of normal cognition. For example, general anesthesia acts on the GABA and NMDA receptors interrupting the functioning of CTS followed by a loss of consciousness. Despite the long history of this medical procedure, the exact chemical mechanism still remains a mystery. But one thing is clear: intervention in the dynamics of oscillatory processes causes disharmony in the network.

One of the most common substances used for general anesthesia is propofol. The supposed mechanism of action is the modulation of GABA receptors and slowing of the ion channels closing time or even blocking the sodium channels. One EEG study showed that propofol causes a major reduction in the brain's information integration capacity (Lee et al., 2009). Another study used fMRI to look at the spatiotemporal organization of connectivity in cortical and subcortical regions under the influence of propofol. The researchers observed a breakdown of subcortical-cortical and corticocortical connectivity, especially in CTS, and a reduction in whole-brain spatiotemporal integration (Schroter et al., 2012). The recorded activity of individual neurons in the brains of patients during surgery under anesthesia with propofol showed that the loss of consciousness was associated with a desynch of oscillations leading to the activity becoming fragmented in time and decorrelated in space (Lewis et al., 2012). It is important to emphasize that a large doze of propofol causes such a disruption of brain activity that it is used as a lethal substance in executions and assisted death.

In the case of pharmacological intervention, there is no disruption of the neuronal anatomic connections and overall functional connectivity is undermined because of desynch in the system. But in some pathologies leading to a massive loss of connections in the CTS or in case of lesions to the substrate the loss of consciousness, coma or vegetative state is the outcome. Both temporal and spatial disruptions lead to consciousness and personality fading away thus confirming that CTS is the physiological, physical and technological substrate of phenomenal consciousness (subjective experience of the world and of oneself in this world).

We have to note that although the thalamus is a complex connecting and computational hub at the center of informational flows in CTS its ability is limited and our immediate conscious awareness is a thin layer of overall processes that we combine in the word Mind. Most of them are subliminal and experienced by us as "automatic." Only salient, new and sufficiently slow signals reach the level of higher cognitive processing that requires time and energy. This may also account for the limited scope of our selective attention. We cannot focus on several tasks at once. We can only switch our attention, transferring the rest of the signals to the background processing mode. This is not only an experience we get from our daily life but an experimentally confirmed fact. For example, in a dual-task setting conscious processing of a first target causes a bottleneck on the routing of a subsequent target (Pashler, 1984). The result is that the second target processing is either postponed till the first target processing finishes ("psychological refractory period") or there is no conscious processing of the second target at all ("attentional blink"). Both may occur within the same experiment (Marti et al., 2012). During the attentional blink, a mildly masked stimulus, normally quite visible, becomes invisible when attention is diverted to another task (Sergent et al., 2005). Numerous studies show that subjects intensely engaged in mental activity fail to notice salient but irrelevant sensory stimuli. The phenomenon is referred to as "inattentional blindness" (Mack, Rock, 1998).

But the slowness of the conscious process is compensated by the range of the created reality model that provides strategic advantages for decision-making. This concerns all modalities but we can take vision to illustrate the idea. For example, there is a condition called "blindsight." This term seems paradoxical if we proceed from the intuition that our vision is about creating visual images. But this is only part of the truth. The first studies of brain injuries during wars showed that the loss of part of the visual field was associated with the location of damage to the visual cortex, and complete lesions led to the loss of visual images (Lister and Holmes, 1916; Holmes, 1918). This topographical correspondence was not particularly astonishing. But when researchers discovered that these people could react to what was happening in the world as if they could see, surprise was inevitable. Many even started talking of some extrasensory way of knowing the world. Later studies of brain activity revealed the actual state of affairs.

Visual perception starts from the eyes but apperception takes many ways and involves many networks. Some of them do not involve cortical areas. They are phylogenetically older pathways used by the species that do not have a cortex at all. So, a person with a damaged visual cortex is not blind but cortically blind. It means that the brain is not producing visual representations as images of the world.

We can only speculate about what picture of the world animals with no cortex have. But when it comes to humans we can judge by their actions and compare them with self-reports. For example, a person with a damaged visual cortex says that he can't see anything, but raises his hands to block an object flying in his face. It means that he can see. But when asked, he replies that he did not see the object and the action was involuntary. When asked to forget about the lack of vision and try to reach objects in the field of vision, a person does this with not perfect but still amazing accuracy. People with blindsight can even differentiate shapes and colors (Ajina, Bridge, 2016). They usually explain it by good guessing but experiments show that the accuracy is far beyond a lucky guess.

Blindsight was surprising for the patients themselves, who insisted that they do not see what they see, and for the researchers, who did not understand how can patients see without seeing. The reason is that we are cortex-dependent in our perception of the world and in our understanding of the perception. Even with all the evidence about non-cortical processing researchers insisted that the residual vision in blindsight is due to some part of the cortex still being involved in vision even in case of total visual cortex loss. Indeed, as we have discussed in the previous part of the study, the cortex is a universal processor and its zones are highly interchangeable. Still, this does not answer the question of how other species see while lacking cortex as a structure in principle.

If we consider blindsight from a technological perspective, it means that the full cycle of the algorithm is disrupted due to the disappearance of one of the elements — the visual cortex. Thalamus keeps on receiving input from the retina but no representations of the incoming signals are created on the awareness level. For instance, a lesion of the V4 area of the visual cortex can destroy conscious color perception in the contralateral hemifield (Zeki, 1993). But only lesions to the retina and optical tract lead to total blindness. There is no "extrasensory perception" for a simple reason: if there are no sensory receptors and/or their connections to the processing filters, signals cannot be encoded. Even non-eye-related processing of light signals is still performed by photoreceptors (for example, skin receptors of cephalopods).

The conceptual reason is gradually evaporating so blindsight is not a surprise anymore. But the physical reason will stay with us. In evolution, the processing of signals by the cortex structures started to dominate in creating the reality model and will continue to do so as long as there will be species with this part of the brain. Yes, this processing takes longer time and requires energy, but it gives these species advantages that are hard to throw out. "Wastebins" of evolution were reserved for branches that did not have an adaptive advantage. But what worked well was well preserved. The addition of structures does not cancel out the old structures.

Corticalisation of functions leads to situations when lesions to the cortex or pathological processes in it disrupt normal conscious processing and undermine adaptability. But it does not mean that species that do not possess such structures are invalids. The CTS hypothesis, obviously, applies only to animals that possess such a system. There is abundant evidence from normal and pathological functioning of the CTS that it is the generator of the conscious content of the Mind. But other species have structures that perform the same function of modulation and integration, thus producing representations and the reality model with a sufficient level of adaptivity. They obviously have bodily agency and intentional agency.

The self-reflective and narrative agency level is the prerogative of a fully and normally developed human CTS. The marker of such an agency is the ability of a person to provide a subjective report of events. In case of the CTS functioning disruption this ability diminishes or even disappears. Cases of blindsight clearly show that such people cannot report the visual experience as their brain does not produce conscious visual images.

But it does not mean that a frog or a fish have only blindsight and are not aware of what they see, otherwise they would not survive. Even such simple creatures as lampreys have brain structures that can produce complex adaptive behavior based on vision and other senses. We cannot be sure about the level of their awareness, as we cannot get their subjective reports. But then again, they cannot produce them for the simple reason of not having brain structures that are responsible for the self-reflective and narrative agency. In this sense, their processes are unconscious.

The issue of conscious and subconscious (subliminal) levels of the Mind has a long history. The term subconscious was coined in 1889 by psychologist Pierre Janet in his doctorate of letters thesis "De l'Automatisme Psychologique." He proposed that underneath the layers of thought, there is a vast subconscious mind that acts automatically, i.e., without awareness (hence the title of the paper). Later, this term has been used widely by various psychological models. Still, without sufficient knowledge of the underlying processes in the substrate, they remained highly speculative. To compensate for this obvious deficiency, the search for neural correlates of consciousness began.

But in studying physiology we must not forget that it is the embodiment of physical processes that employ a physical mechanism. Thus, any question about the functioning of the Mind on any level and in any state has to be considered from four sides: psychology, physiology, physics and technology. The enumeration order does not mean the priority order. It is just a historically based order. First, we studied the external manifestations (mental activity and behavior). Second, we looked into the physiological substrate that produces these manifestations. Next, we should explain the physical mechanism behind the physiological processes and the technology of its implementation in the substrate.

If we do not define the objects and subjects of our study in physical terms, we cannot fill the explanatory gap between psychology and physiology. If we do not posit questions in physical terms, we cannot get physical answers. This is the major problem. It concerns all aspects of the Mind, including the difference between conscious and subconscious processing which is phenomenally sometimes obvious and sometimes not.

The term subliminal etymologically means "below the threshold" (from Latin limen — threshold). However, the nature of this threshold remained unclear. Again, psychological theories gave a variety of interpretations, not based in any way on real processes in the substrate. It seemed that the development of experimental technologies would clarify this issue. But currently, we witness an accumulation of data that is either not taken into account by any model or contradicts previously proposed models.

The authors of one overview described the state of affairs: "What brain mechanisms underlie our capacity to become aware of a specific piece of information, while many others remain non-conscious? ... Visual illusions and a great variety of other normal and pathophysiological conditions such as sleep, anesthesia, blindsight or hemineglect provided empirical windows into this phenomenon, by providing minimal contrasts between conscious and non-conscious brain states. Here we review the recent advances made possible by this contrastive approach. We specifically focus on how these findings inform present-day theories of conscious processing. At present, there is no accepted computational theory of this function. Our hope is that the present review may point to the key ingredients that will lead to one" (Dehaene et al., 2014).

They review many studies that show the following results: cortical areas are activated when the stimulus is reported by the subjects as unseen; recognition of the identity of pictures, words, faces, symbols and their meaning can happen without conscious awareness; attention is deployed and enhances processing even if its target remains non-conscious; attending to a stimulus and becoming conscious of it have distinct signatures that occur on distinct trials and at different times; a non-consciously processed visual cue can trigger inhibitory motor control circuits; error-detection and task switching can happen non-consciously; changes in gamma band power can be evoked by stimuli that remain outside of conscious processing; integration of a complex visual scene can happen subliminally.

These findings refute the following long-standing ideas: non-conscious processing happens only at an early perceptual level; attention and executive control are conscious processes; information integration is the marker of consciousness; specific processors or specific frequencies are the seats of consciousness.

What remains intact is the idea that there is some kind of threshold that separates subliminal and supraliminal levels. It is not an anatomical one as the classics of psychology and many physiologists thought. We cannot say that cortex is the seat of consciousness and subcortical structures are the non-conscious domain. We cannot say that it concerns a specific frequency band or some integrated information measure as some recent models suggest. There seems to be no order in the brain that would explain the phenomenology of conscious and unconscious states that we experience in our mental life. But it just means that there is no order in our models. The brain's got an order.

The authors of the review are proponents of the Global Workspace Theory (GWT) that "proposes that conscious access stems from a cognitive architecture with an evolved function of the flexible sharing of information throughout the cortex. While non-conscious stimuli are processed in parallel by specialized cortical processors, conscious perception would be needed in order to flexibly route a selected stimulus through a series of non-routine information processing stages. Global information sharing and routing would rely on a set of interconnected high-level cortical regions forming a 'global workspace' and involving primarily the dorsolateral prefrontal cortex, but also additional hubs in inferior parietal cortex, mid-temporal cortex, and precuneus, and now described as forming a 'rich club' network … Conscious access would occur when a piece of information enters a distributed network of cortical areas tightly interconnected by long-distance axons" (Ibid).

The model speaks only about anatomic areas of the brain and their connections. The time domain is absent. The focus is on the space domain. This is reflected in the name "workspace." But the major problem is that the theory again tries to "jump over" the explanatory gap between the physiology of the brain and mental phenomenology. It totally ignores physical and technological issues. It sounds as if all the questions of how information is created, transmitted, integrated, stored, retrieved, and broadcast have been solved by the model. But in fact, the authors of the GWT omit all those issues and go directly to proposing that everything happens in a global workspace of well-connected "rich club" members. This is considered to be the main achievement of the model.

The trick is that it is impossible to refute the model as it speaks of general ideas which are about everything and nothing in particular. Even enumerating certain areas of the cortex as part of the global workspace does not add any specifics, as the authors themselves note that "findings support the view that virtually any cerebral processor may operate in a non-conscious mode." So, it is not only space parameters that define the workspace. Stating the old fact that there are specialized and global processors in the cortex does not add anything to our knowledge. Jumping over the gap does not help, as we keep on falling into the abyss of the unexplainable. We need a bridge. We need to answer the physical question "What?" and the technological "How?"

What is information physically? How is it created? How is encoding performed? What is the physical nature of the neural code? How do the cortical regions perform routine and non-routine information processing? What is this process physically? How do they flexibly route? What is the physics of the transmission channels and process? How do high-level cortical regions integrate global information and broadcast it? What is the integration process physically? How do they store information and share it? What is the physics of memory storage and retrieval? We have been dealing with all these questions in the previous parts of the study. Here we can only point out that these questions concern both conscious and non-conscious levels of the Mind.

As for the threshold between these levels, we can find hints at some answers in the empirical studies taken by the authors for the review. We should only apply to them a comprehensive model. The authors note key ingredients of the empirical findings: "Recurrent thalamo-cortical networks provide a simple and generic implementation of elementary stimulus categorization processes. Recurrent NMDA connections impose slow accumulation dynamics and multi-stable 'all-or-none' behavior, whereby the incoming evidence either quickly dies out (corresponding to subliminal processing) or is accumulated and amplified non-linearly into a full-blown state of high-level activity. This global 'ignition' has been proposed as a marker of conscious perception. Indeed, empirically, when stimulus strength is varied, the early stages of non-conscious processing typically show a linear variation in activation, whereas conscious access is often characterized by a late non-linear amplification of activation which invades a distributed set of parietal, prefrontal and cingulate areas ... In EEG, MEG, and intracranial recordings, conscious stimuli, compared to matched non-conscious ones, induce a late (~300 ms) and sudden increase in slow event-related potentials, gamma power and long-range beta and gamma synchrony ... Thus, whether a stimulus is detected seemed to be determined by an accumulation of pre-stimulus bias ('prior') and stimulus-evoked activation ('evidence')" (Ibid).

Let's translate these findings into the language of TTT hypotheses and predictions. The experiments show that conscious processing involves CTS that includes filters-modulators, and preliminary, intermediate and higher filters-integrators. The system functions using an iterative PAAL algorithm with recurrent flows of projection (prior) and introjection (evidence). For both flows to reach the consciousness level there have to be certain thresholds reached that include such parameters as amplitude, cross-frequency interaction and phase coupling. It is a non-linear process of activation-inhibition interplay that leads to the synchronization of wave structures as representations of the higher cognitive level that we experience as conscious awareness. To find the threshold between the conscious and unconscious levels of the Mind we need to evaluate all of the above parameters.

This is a short description of the physics and technology of the "ignition" of the spark of conscious processing performed by the CTS. It clearly shows that both spatial and temporal parameters are important and we cannot rule out or even downplay any of them. The "rich club" members that form conscious awareness and the feeling of integral personality are connected in time and space. We need to understand how these connections facilitate the physics and technology of the information-creating process that we call the Mind.

CHAPTER 5

THE SECRET OF THE UNIFIED SELF

It is a convenient and exceedingly useful invention on the part of the brain. It binds, therefore I am!

Rodolpho Llinas

How does the brain create a coherent model of reality and what we perceive as the unity of Self while maintaining the representations' identity and at the same time preventing the picture from falling apart into separate pieces? This binding problem can be formulated using the musical analogy. How does the brain create the unified polyphonic and polyrhythmic music of the Mind? As we have noted many times earlier, this analogy helps us to put the question in a physical perspective and leads us to the physical answer as a necessary bridge between physiology and psychology.

The neuroscientist Horace Barlow wrote 60 years ago: "A wing would be a most mystifying structure if one did not know that birds flew. One might observe that it could be extended a considerable distance, that it had a smooth covering of feathers with conspicuous markings, that it was operated by powerful muscles, and that strength and lightness were prominent features of its construction. These are important facts, but by themselves they do not tell us that birds fly. Yet without knowing this, and without understanding something of the principles of flight, a more detailed examination of the wing itself would probably be unrewarding. I think that we may be at an analogous point in our understanding of the sensory side of the central nervous system. We have got our first batch of facts from the anatomical, neurophysiological, and psychophysical study of sensation and perception, and now we need ideas about what operations are performed by the various structures we have examined ... As with the bird's wing, the summaries are in physical rather than biological language ... A bird's ability to fly is certainly an important fact, but it might easily be missed by someone concentrating his attention too narrowly on the anatomy and physiology of wings" (Barlow, 1961).

We should look at anatomy from physical and technological perspectives. Otherwise, the accumulation of details of physiological knowledge is not only unrewarding but, in some ways, even plays a negative role. "Bricks" accumulate at the construction site, but the builders do not know what to do with them since there is no architectural project (model). The "brick factory" of neuroscience operates at full capacity, and the construction site has already turned into a heap of material. We must make sense of the accumulated treasure.

If the function is binding, then we need to ask ourselves about the physical mechanism that can provide the possibility of binding. Next, we should look at how the specific physiological substrate uses this mechanism to perform the binding function. This is the only way to build the bridge to cover an explanatory gap concerning any psychological function including, probably, the most important one: the unity of Self.

Here it makes sense to repeat basic assumptions of the TTT about the functioning of the brain:
1. The hybrid analog-discrete-analog code.
2. The wave nature of representations.
3. Interaction of the system's elements via synchronization.
4. PAAL algorithm.

If we look at them from the binding problem perspective, we can see that each item is actually one of the answers to the question. The analog-discrete-analog conversion is essentially about differentiation and integration. Wave physics and sync are the underlying mechanisms behind the coding technology. The iterative self-referential algorithm binds all the processes together.

Let's see if the overall anatomy of the neural connections corresponds to the task performed by the physical mechanisms and technological solutions. In the brain, there is no direct connection of all neurons with each other. Most of the neurons are connected with their nearest neighbors, and long connections make up only a few percent of the total. How is the general binding ensured with seemingly sparse system-wide connections? It turns out that this topology of connections is, in fact, optimal. Such a structure is called a small-world network (SWN). The diagrams below show different types of networks:

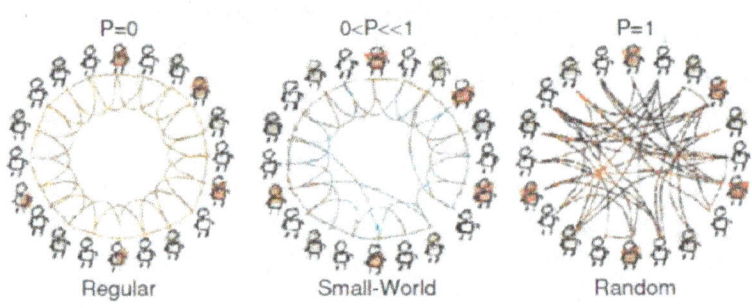

Watts, Strogatz, 1998

Clusters (human figures) reflect the local structure, and the average length of the chain from one cluster to another (degree of separation, "the number of handshakes") reflects the global configuration. Random networks have a low degree of separation but are poorly structured by cluster. Regular networks are highly structured, but clusters are highly separated. The optimal balance of two variables (clustering and separation), which creates an effective information transfer structure, arises in networks organized according to the small-world principle.

The Small-World Theory developed as a combination of different ideas. For example, "Pareto principle." Wilfredo Pareto, a sociologist and economist, when studying the incomes of Italian households in 1897, concluded that 20% of the efforts yield 80% of the result. Later this pattern was observed in many areas from the economy (20% of people have 80% of the capital, 20% of customers bring 80% of the profit) to the Internet (80% of users visit 20% of sites). Observations show that biological systems follow this rule. For example, the size distribution of trees in a forest is the same (Farrior et al., 2016). Of course, the numbers are approximate, but they show the meaning of the rule: the organization of the process in complex systems has a regularity, which is sometimes called the power law.

Different authors formulated this rule in their way. Still, the essence came down to one thought: self-organization leads to the emergence of systems that increase the efficiency of energy conversion and transmission. Some called it the fourth law of thermodynamics. Others called it the law of evolution since it said that living systems that maximize the efficiency of energy and information flow survive and thrive. This is a purely phenomenological rule, stating the observed pattern but not explaining the mechanism for increasing efficiency.

The current interest in such patterns has been revived in studies of the global Internet network (Barabasi et al., 1999). When looking at the topology of this network, researchers found that some nodes have much more connections than others. They were called the "hub node," in the case of the predominance of outgoing communications, and the "authority node," in the predominance of incoming communications. Sites on the Internet are distributed according to the power-law: some sites are central in the sense of their connection with all others directly and indirectly. And this is not a conspiracy but a product of the nonlinear dynamics of network evolution. This is how the self-organization of a complex scale-free system occurs. There is no definite uniform scale, but there are patterns of relationships and their hierarchy.

This distribution of ties is also called the heavy-tailed distribution. Albert Barabasi and colleagues proposed an explanation of the mechanism of such distribution and called it "preferential attachment." Simply put, it means that those who came earlier have preferential connections because they had more time for their establishment. As in an elite club: newcomers are marginalized at first and prefer to build a relationship with those who have been there for a long time and are a hub node or authority node, or both at the same time. A beginner is not in that position, but time can change everything if the system is dynamic and scale-free, i.e., not rigidly hierarchical. Democracy of a dynamic structure does not mean anarchy and lack of hierarchy, but the potential for turnover of power, i.e., adaptive system evolution.

In mathematics, there is a random networks theory. In short, it is about how, independent of the number of nodes in the network, a small percentage of randomly located connections is always sufficient to connect the network to a more or less integral state. Example: 300 nodes have 50,000 variations of their connection with each other, but 2% of random links are sufficient to make the system fully integrated, in the sense that elements will be connected directly or indirectly. Here is a paradoxical name again: random network. Observations show that it's not random at all. There are patterns in this process, and they are subject to forecast.

The meaning was well formulated by the title of the article by the sociologist Mark Granovetter: "The strength of weak ties" (Granovetter, 1973). Strong ties ("handshake" distance) form nodes (family, clan, circle of friends, colleagues, etc.). Weak connections between nodes create the effect of the overall unity of the system. They act as bridges between clusters of strong ties. Even the small number of long bonds makes a substantial binding weight of the entire system. The system remains orderly but also dynamic ("randomly connected").

Long-distance communications reduce the degree of separation. They work as "handshakes" between distant cluster nodes. If we represent an ordered hierarchical structure, then one of the archetypal images will, of course, be a ladder. It has a right to exist, but it is limited in its effectiveness. If there are a lot of stairs, you need connections between them. In an ordered linear structure (regular network in the Watts-Strogatz model), movement from one element to another takes a lot of time. When a structure outgrows a certain level, such a path becomes ineffective. For signals to spread across the network quickly and efficiently, it inevitably moves from a linear to a local-global organization of the small-world type. The picture below shows developing SWN:

It is no coincidence that the evolution of human society has led to globalization with a paradoxical effect: the big world has become a small village where you sneeze at one end, and they say "bless you" at another. This, of course, has its drawbacks: the rapid spread of viruses, economic crises, and other troubles. You have to pay for everything.

In the same way, the development of multicellular organisms as cell societies was accompanied by an increase in the efficiency of the internal flow of energy and information and increasing dependence of everyone upon everyone. The brain structure of higher animals demonstrates the advantages and disadvantages of such a network vividly. But the benefits outweigh, so we have what we have: a neural network with tens of billions of elements in one head, where each neuron from head to heels is several "handshakes" from any other.

Strogatz and his colleagues took as an object of research the power network of the western states of the USA, uniting 5,000 power plants. It turned out that it was very close in terms of the degree of separation to random, but it was also with a high level of clustering (16 times more than for random).

Pay attention to the logic of events. Based on a theoretical model, mathematicians calculated and simulated on a computer how an optimal network should be organized. Then they took an actual complex network and calculated its performance. It turned out that a real network, created by the evolution of engineering practice of increasing efficiency by trial and error, has laws that can be expressed mathematically. It is not that mathematicians fitted the results to reality. Quite the opposite: a complex system (in this case, energy network) in the course of evolution acquired certain patterns of optimal organization, which can be expressed mathematically.

Then the researchers took the living system: one of the simplest and most studied — the Caenorhabditis elegans worm. It consists of 959 cells, and its size is about 1 mm. Its genome was calculated back in 1998. It served as a model for the study of many processes: from the transmission of information to the death of cells. Its nervous system of 302 neurons is mapped in detail, and it can be downloaded to a computer. Scientists evaluated the state of connections in the network. Although 118 types can be distinguished among 302 neurons, it is crucial for network analysis that if two neurons are connected by a synapse (electrical or chemical), then these are two network nodes. As a result of the investigation, the system turned out to be organized by the small-world principle. The average chain length (degree of separation) was only 18% greater than for random, and clustering was six times greater.

The authors' hypothesis, as a conclusion of research: clustering reflects the structure of the nervous system, the populations and their connections, the level of specialization; chain length reflects the efficiency and speed of information exchange at the level of the entire system.

Strogatz wrote: "Two radically different networks, the power grid and the nervous system: one created by mankind, the other by evolution. One is among the largest machines ever built, a sprawling web of synchronized generators linked by hundreds of thousands of miles of cable. The other is a microscopic filigree, the product of millions of years of natural selection, a lacework snuggled in the body of a worm. And yet despite all their differences, their architecture is strikingly similar. Both networks are almost as small as they could possibly be. Both are highly structured and definitely not random" (Strogatz, 2003).

Strogatz and his colleagues continued the study and took a social network. At that time, the Kevin Bacon Game was very popular: a network that connected all the actors in the world based on their acquaintance with Kevin Bacon. It began when this Hollywood actor in 1994 half-seriously half-jokingly said that he worked with everyone in Hollywood directly and indirectly. A network of "six handshakes" of actors was created based on data from studios. The calculations showed that the degree of separation (the length of the chain from any actor on Earth in all history of the movie industry to Bacon) was 3.65. For a random network of this size, it would be 2.99, i.e., the smallness of this world was close to minimal. But clustering was very high: 0.79 (3,000 times more than for a random network). This is how the social network self-organized into optimal parameters for effective communication and interaction.

Just imagine organizing the exchange of information between billions of brain cells. To connect everyone with everyone linearly, an incredible number of connections would be required. The more "wires," the more significant is the loss of energy-information, the loss of time, the more costs, the greater the risk of errors, faulty connections, etc. In the end, it is simply inefficient. Furthermore, there is a limit for extensive development: the "box" of the skull is not unlimited. The way out is to develop an optimal network structure.

The neuron connections look very confusing from the outside, but in this "mess" lies the efficient organization of the network. Studies show that the volume of links is much smaller than if everyone were connected with everyone. This sparse connectivity can be confusing if we do not consider the network topology and efficiency. The secret is simple: the small-world network saves spatial, temporal and energy resources. It also allows effective self-management.

Imagine a system of many independent elements that must solve their "personal" problems and global ones. The elements themselves do not see the global picture. There is not one single element that looks at everything from above, knows what everyone is doing, and resolves the general situation. In computer technology, organizing simple elements into a complex system is called the problem of density classification of one-dimensional binary automata. It sounds complicated, but the point is to create network rules for making any decision, even the basic binary: Yes/No.

There are many elements in the network, and each one has a voice. But we need a consensus; otherwise, the system does not work. It is a democracy of electronics. Each element can take one of the options for this binary solution. But how can we solve the global problem with the local voice of each element? What about the SWN scheme?

Calculations showed that when a certain threshold of random long connections was overcome, the majority rule worked with an efficiency of 88%. It is the math of democracy. The network spontaneously became capable of making decisions (in this case, calculate) if an effective communication structure was established. Strogatz wrote about it this way: "A dumb rule (majority rule) running on a smart architecture (a small world) achieved performances that broke the world record" (Strogatz, 2003).

In a complex system, there is no alternative but democracy, as an optimal balance of locality and globality, clustering and chain lengths, structure and freedom, determinism and indeterminism. If we learn from the experience of the world around us and from our own experience, which shows that any shift in balance to the extreme of a rigid structure or to the chaos of anarchy is simply not effective, maladaptive, doomed to failure, then we will approach a more optimal social system. We just need to remember that this is how our brains work and learn from our brains — a paradoxical but not absurd task.

Unfortunately, the principle of SWN has its downside. We are talking, for example, about the spread of viruses (of any kind — pathogenic, informational, ideological, economic, etc.). They demonstrate the principle of energy-information distribution in the SWN. The virus first appears in a small cluster but then begins to jump over long bridges of long-distance connections. Thus, the AIDS virus, having begun in small tribes of Africa, quickly spread thanks to transcontinental flights. The latest epidemics of different types of influenza and coronavirus began in one place and almost instantly covered the whole world thanks to literal handshakes of people constantly moving around the world. From the point of view of viruses, this is a good efficiency. After all, they have their point of view, their own self-organization.

Humanity periodically experiences outbreaks of viral diseases, and all of them occur according to one scenario. And the better the connections are in the global "village," the faster the viruses spread. The obvious way to deal is by interrupting global communications and clustering (isolation). But the system either develops and, therefore, increases the speed and effectiveness of connections, or falls apart, i.e., ceases to be a system, dies as a whole. We can, of course, temporarily switch to self-isolation, but the question arises: what is temporary? How long is the isolation? Until the virus subsides? But where is the guarantee that another one will not follow? There is a saying: nothing is more permanent than a temporary solution. But the permanent interruption of connections means the death of the system.

As our world becomes smaller and smaller (in the sense of the SWN principle), the risks of the spread of phenomena that are not very positive also increase. Sneeze at one end of the world leads to cough at the other, in literal or figurative meaning. The only thing we can do is study the laws of processes and try to correct them. Manage them so that they do not control us. But at the same time, it is impossible to reverse the process of globalization. And we must remember that this process also has a good side: it aims to increase the efficiency of energy-information exchange.

Even our language is organized according to the SWN principle. For example, studies of the English language structure from this point of view led to the following results. If we take the British National Corpus, as a collection of texts of 100 million words, and examine the links, it turns out that the average chain from any word to any (a link, in this case, means the arrangement of words together) is 2.67. This does not sound surprising. After all, one can say that the variability of words is close to random. But clustering was 4,000 times larger than for a random network (Ibid).

Another example: if you take an old TV with its many parts and wires, or a new one consisting of microchips, all the elements will be in communication with each other, but with a high degree of clustering, because they are compiled in modules. How did it happen? Natural selection, the evolution of electrical engineering. This design turned out to be effective, balanced between reliability, speed and economy.

Any road network would be a simple and illustrative example. A village may have one main road and associated secondary roads. They may not be related to each other since the simplicity and small size of the network allow quick access from one to the other through the central one. With the growth of the network, branches arise, and an optimal communication organization between them is required. In the metropolis, roads are built by the small-world principle, although road engineers may not know about this term and use different words. There are clusters of local roads and highways connecting them. Engineers follow a natural path that dictates them the pattern of development of a living, dynamic system of the city. In the case of normal development, a balanced evolution of all types of bonds occurs. Otherwise, the city will begin to "suffocate" in systemic traffic jams. This is a disease, a pathology of the state of the system.

All small-world networks are resistant to perturbations. But a targeted attack can lead to a chaotic state. Network elements are interchangeable, and the system has plasticity. But a cardinal disruption of connections leads to desynchronization and decay. The influence of the nodes most connected with all elements is strong. These clusters become vital to the survival of the system. For example, the most related protein clusters become the most important for cell function. Studies by Barabasi and colleagues showed that removing proteins with a not very high link rate (this is 93% of all proteins) led to cell death in only 21% of cases. Removal of the well-connected "kingpins" (top 1% of proteins) was fatal in 62% of cases (Ibid).

Freedom is not anarchy-chaos, and there must be a structure in the system; otherwise, it is not a system. The truth of democracy of life is that "all are equal, but some are more equal." Neither economy, nor politics, nor organization of society as a whole, nor the organization of living systems can do without structure. The most effective systems are those with plasticity, interchangeability of elements, the width and depth of their connections ("one for all and all for one"). Such a system has a dynamic hierarchy of connections, where some clusters are well-connected. The flip side of this process is the vulnerability of the system when these high-level clusters are attacked. But here, the saving principle works: if most of the elements are interchangeable, then the system's stability is high.

The robustness of such a network can be demonstrated with an interesting example. A group of researchers experimented with springtails Folsomia Candida. They settled these animals in different jars connected by plastic tubes (Gilarranz et al., 2017). First, they studied the dynamics of individual clusters, the processes of their reproduction and mortality. The populations grew, but then they reached the limit and stabilized. Then the researchers staged a local catastrophe: the complete removal of the population in one of the clusters. Naturally, springtails from neighboring clusters began to occupy free territory. But scientists removed all new arrivals. It was an "open wound" where the living resources of the entire colony flowed away. But after a while, there was a buffer effect. The populations of the nearest nodes were most affected, but the process stabilized. The further the node was from the "catastrophe," the less it was affected. The buffering effect was twice as large in the population separated by two tubes than the one nearest to it. Scientists made a computer simulation of the situation, but with the participation of thousands of nodes, variations in population density and perturbation strength. The results were similar. The increase in complexity led to even stronger resistance to perturbation.

Usually the system evolves smoothly, and the nonlinearity of the dynamics leads to a change in structure (a shift in the centers of power, "re-election") as a guarantee of stability. Often such an intuitive factor of system stability is ignored in the process of political self-organization of society. To the extent that it is considered the other way around: the dynamics of power, re-elections, a change of leadership are supposedly a sign of instability, and society begins to freeze around the "stable vertical of power." However, the internal natural instability of such a design sooner or later leads to destruction. Even a minor attack on a narrow center can lead to disaster for the entire system. Nature has come to the efficient organization of living systems by trial and error over billions of years. Human society develops for dozens, maximum hundreds of thousands of years. This is just a moment of the evolution of the Mind, and social organization is just beginning to approach effective complex structures like our brain.

The possibility of creating connections according to the SWN principle does not mean automatic implementation. Theoretically, the street sweeper is connected with the president of the country via six handshakes. But the question is how to find the right connections so as not to shake hands with billions (an unrealistic and ineffective way of reaching out to the president). The living system works on this task during evolution in phylogenesis and ontogenesis. Each cell of the body needs to be within a few "handshakes" from the central nodes of the nervous system.

In the development of a neural network, a living system must maintain a balance. It is necessary to build up connections (both short and long), but not higher than the level required at the moment. Otherwise, it will become energetically and informationally ineffective. And it needs to increase the capacity of the central nodes but not overload them. Otherwise, there is a risk of "traffic jams" in the signal flow and an increase in dependence on these nodes, which may be subject to internal and external negative influences.

The development of the SWN type networks was an ingenious course of evolution. It is the result of not only necessity but also freedom. For example, during the development of the neural network, a process of neurotropism occurs: a phenomenon of a directing effect on the growth of nerve fiber by various chemical or physical factors. The end of the nerve cell (growth cone) can actively move along the substrate, has a high sensitivity to multiple chemicals, and is equipped with dynamically changing projections (filopodia). If you change the target's location, the cone changes the direction of growth and finds the desired cell. How? Neurons communicate not only after the establishment of direct synaptic connections but before that. They send chemical signals to each other (neurotrophins NGF, BDNF, NT3, NT4/5), attracting other neurons to grow in a specific direction (chemotropism).

This is the establishment of "family ties." The families and clans of neurons are formed, where the connections are very close (literally). However, there is not much freedom in such relations. In the process of evolutionary phylogenesis of the nervous systems, ties first grew on such a basis. If we look at the stages of the development of the human brain in its ontogenesis, the same thing happens. First, the closest "family ties" are established. But as they develop, the distribution of ties according to the SWN principle and the growth of the system's efficiency are underway.

Chemical signals set the direction in the local dimension, but as the space of the system expands, their commands become less deterministic, and neurons gain more freedom. The specificity of contact formation is not absolute for all processes of nerve cells. Genetic and epigenetic factors work. In addition to chemotropism, growth is influenced by numerous other factors (lines of mechanical stress, gradients of energy processes, processes in glial cells, etc.). Experiments in which different directing factors were presented to the growing process showed that this is a process with many degrees of freedom. It is not a linear single factor scheme but the distribution of all factors in the spatial and temporal dimensions.

In precisely the same way it happened with humanity as a social network, as a living system consisting of individual elements-individuals. We left the clan social system. We gradually leave paternalistic, autocratic systems built on the principle of "one over many." The whole history of the development of social structures can be called the history of expanding individual freedom. It can be assumed that, provided that humanity is preserved as a species, the further evolution of the social structure will go in the same direction of a parallel increase in the level of individuality and growth of complexity of the entire system. This is indicated not by some political doctrine but by the logic of the evolution of living multicellular systems as cell societies, including the development of the nervous system as a whole and the brain in particular.

Family, tribal, national ties remain, but society has qualitatively changed due to an increase in the degree of freedom of the individual and the possibility of expanding relations. We are becoming a "big village," where a neighbor can be at the other end of the globe. The creation of technologies for rapid movement in space, the emergence of information networks based on the same principle of SWN allowing us to "shake hands" with a person thousands of kilometers away is not an accident but a pattern. Each one of billions is in direct accessibility to another precisely due to freedom.

In phylogenesis, the stages of maturation and growth of the brain network in humans, as a species, coincided with qualitative leaps in the development of culture. The Upper Paleolithic era, when the modern Homo Sapiens spread throughout the Earth, was marked by a jump in tools' level and the formation of more complex social structures than other primates had. In a small group in which modern primates live and early people lived, interaction is determined by daily practice and close but short ties. Such social structures are both linear and rigid. In more developed systems, connections have recursive logic. A hierarchical recursive relationship with a tree structure can be described as follows: not more than one to many. It is fundamentally different from a one-over-many structure logic, which leads to the linearity of a rigid vertical.

In an effectively organized complex structure, each element is located only a few "points" from any other. The amount of information and the effectiveness of its transmission is huge. These are signs of the emergence of a small-world network. The increase in the system's volume leads to the fact that a simple form of face-to-face interaction becomes problematic. The growth of each element's individual level of development means, accordingly, the uniqueness of interpersonal relationships. The growth of the network and the development of each individual, as its element, required more diverse connections to maintain social coherence. As a result, society acquires a qualitatively different level.

Modern humankind is informationally an SWN. The paradox is that people in one part of the world, both culturally, intellectually, and emotionally, can be closer to people on the other side of the world than people in a neighboring apartment or house. This situation is completely new in history. But management in some societies remains at the level of clan hierarchy, striving for the extreme of complete regularity and the linear vertical of power. Such rigid structures have long been ineffective and lead from one collapse to another. In a short perspective, the illusion is created that such a management system works. But from a historical perspective, it becomes clear that previously such structures could exist for centuries, and now they are collapsing after several decades and even years.

We can draw an analogy with economic structures. "Manual control" is possible at the small business stage, when one can be the owner, the director, the manager of all the links down to the basic, and even the executor. But if a business develops into a corporation, then inevitably comes not only the multiplication of owners and levels of structure but also a radical change in the management paradigm. If the "founding father" continues to pull all the strings, then the collapse is inevitable. But letting go of the reins is also impossible. The level of management should grow in line with the increasing complexity of the system.

Fortunately, the topology of the neuronal network organization lies between extremes of complete regularity (linearity, rigidity) and complete randomness (anarchy, chaos). Thanks to this, it is flexible and can overcome its own drawbacks. An organization based on the SWN principle can dynamically change connections and structure. It is simultaneously regular and variable. Such an organization allowed us to get rid of a relatively narrow, rigid range of the reality model of many previous stages of the evolution of living systems, to expand the horizons of our world.

Hypothesis:

The nervous system is organized as a small-world network. It consists of nodes, which are interconnected by short links and form clusters. These clusters have distributed long links with other nodes. When an optimal balance between clustering and path length is reached, such a network demonstrates high information transfer efficiency, processing power, stability and flexibility.

A small-world principle is a tool for creating a unified and coherent structure. It contributes to the synchronization of the neuronal ensemble, and the synchronization process moves the system towards further development of optimal network organization. Usually, both factors interact in such a cyclical causality for increasing the efficiency of the energy-information exchange. And in the opposite direction for pathologies: a violation in the network topology creates conditions for desynchronization, which in return negatively affects the network.

Since the development of the SWN model, there have been many studies and confirmations that it can be applied to the brain. Here we will just focus on the ones that show that this kind of organization contributes to synchronizing the network into a harmonious ensemble. One study looked at the frequencies of the brain using wavelet decomposition of magnetoencephalographic time series. The authors report on the result: "We found that brain functional networks were characterized by small-world properties at all six wavelet scales considered ... Global topological parameters (path length, clustering) were conserved across scales ... Moreover, the synchronizability of the networks in all scales and states is close to the threshold of 0.01, which marks the lower limit of the transition zone from globally ordered to disordered behavior in systems of coupled oscillators" (Bassett et al., 2006).

First, there is a manifestation of the dynamics of the small-world network, the formation of hubs and nodes, between which fast and effective connections are established with a low degree of separation and a high level of clustering. These dynamics depend on the state of the system. They have stable parameters but also change adaptively in the direction of increasing frequencies and expanding connections during activation and switch to the energy-saving mode at rest (low-frequency and local activity). Second, the proximity to the disordered state was calculated, which turned out to be at the lowest level, i.e., as far from desync as possible. In other words, the synchronization indicators in all ranges and states are generally at the highest level.

Other researchers approached from the reverse side: they created an artificial computer simulation of a neural network of the SWN type and studied the reaction of such a network to different tasks. The authors report the result: "We find that small-world networks perform an order of magnitude better than random ones, enabling reliable discrimination between inputs ... Furthermore, we show that small-world architectures operate at significantly reduced energetic costs and that their memory capacity scales favorably with network size ... Our results suggest that mammalian cortical networks, by virtue of being both small-world and topographically organized, seem particularly well-suited to information processing through polychronization" (Vertes, Duke, 2010). By polychronization, authors mean synchronization in different frequency ranges.

One more team simulated neuronal networks of various topologies and found that "random connectivity topologies give rise to fast system response yet are unable to produce coherent oscillations in the average activity of the network; on the other hand, regular connectivity topologies give rise to coherent oscillations and temporal coding, but in a temporal scale that is not in accordance with fast signal processing. Finally, small-world (SW) connectivity topologies, which fall between random and regular ones, take advantage of the best features of both, giving rise to fast system response with coherent oscillations along with reproducible temporal coding on clusters of neurons" (Lago-Fernández et al., 2000).

Conversely, the studies of the state of the brain functional connectivity in various pathologies showed the disruption of the SWN topology. For example, in one study applied graph theoretical analysis to matrices of functional connectivity in 15 Alzheimer patients and 13 control subjects. The authors reported: "The characteristic path length L was significantly longer in the Alzheimer patients, whereas the cluster coefficient C showed no significant changes ... A longer path length with a relatively preserved cluster coefficient suggests a loss of complexity and a less optimal organization. The present study provides further support for the presence of "small-world" features in functional brain networks and demonstrates that AD is characterized by a loss of small-world network characteristics" (Stam et al., 2007). Another study used fMRI data on functional connectivity between 90 cortical and sub-cortical regions of 31 schizophrenia patients and 31 healthy subjects. Their findings "demonstrated that the brain functional networks had efficient small-world properties in the healthy subjects; whereas these properties were disrupted in the patients with schizophrenia" (Liu et al., 2008).

Studies of the spatial patterns of functional connectivity by computing the synchronization likelihood of EEG during the performance of a working memory task in subjects with schizophrenia and healthy controls showed: "During the working memory task, healthy subjects exhibited small-world properties (a combination of local clustering and high overall integration of the functional networks) in the alpha, beta and gamma bands. These properties were not present in the schizophrenia group" (Micheloyannis et al., 2006). On the other hand, EEG studies of normal working memory functioning showed that distant cortical regions involved in different stages of processing (encoding, retention, retrieval) were synchronized and the specific functional integration of these areas measured by operational synch was unique for each stage of the memory task (Fingelkurts et al., 2003 a,b).

Now here comes the question. What comes first, synchronization or spatial functional connectivity? Sounds like a "chicken and egg" problem. If we look at it this way, we cannot solve it. It is not enough to say that there is a cyclical causality chain. We need to specify the categories that interact within it. For this, we should start with some fundamental premises about the function that this interaction serves.

If we proceed from the assumption that the Mind is a process of creating a coherent reality model, we may look at the above categories from the following perspective: do they contribute to the binding of the various information flows into an integrated structure? The answer seems obvious: they both do. But synch is a universal physical mechanism of energy interactions, and SWN topology is a specific form of network organization that makes these interactions more efficient. From this perspective, synchronization is the primary vehicle that is used by the brain to create a unified model of the world.

The structure of the nervous system reflects the need to harmonize the polyphony and polyrhythm of the Mind. Therefore, studies of functional connectivity not only reveal the properties of the small-world network that contribute to effective communication and integration, but also show the harmonic structure. For example, the authors of one research wrote: "The principle of harmonicity, ubiquitous in nature, also underlies functional cortical organization in the human brain ... This means that transitioning between the functional networks instantiated by the functional harmonics in response to changing task demands is optimally efficient ... Our results point to the existence of the same fundamental principle in multiple aspects of human brain function, including functional integration and segregation" (Glomb et al., 2021).

The authors of one review article noted: "Numerous studies in both animals and humans have shown that synchronized oscillatory activity in various frequency bands is related to a large set of cognitive and sensorimotor functions ... It has been argued that synchronization of neural activity might help to cope with binding problems that occur in distributed architectures ... We believe that the study of synchronization phenomena can lead to a new view on multisensory processing which considers the dynamic interplay of neural populations as a key to crossmodal integration and stipulates the development of new research approaches and experimental strategies. Conversely, the investigation of multisensory interactions might also provide a crucial test bed for further validation of the temporal correlation hypothesis, because task- or percept-related changes of coherence between independent neural sources have hardly been shown in humans so far. In this context, the role of oscillations of different frequencies in crossmodal integration is yet another unexplored issue that future studies will need to address" (Senkowski et al., 2008).

Although temporal coherence studies are not mainstream in neuroscience, they have been going on for a very long time. Nevertheless, the authors of the review are right: the study of the synch of various frequencies is in its infancy. There are many models that suggest that synchronization is an integration mechanism. However, as we have already seen, it is very important to understand what the authors of this or that model mean by the very concept of "synchronization."

For example, the authors of the article "The neuronal basis for consciousness" wrote about the thalamocortical pathways: "The system would function on the basis of temporal coherence. Such coherence would be embodied by the simultaneity of neuronal firing ... In this fashion the time-coherent activity of the specific and non-specific oscillatory inputs, by summing distal and proximal activity in given dendritic elements, would enhance de facto 40 Hz cortical coherence by their multimodal character and in this way would provide one mechanism for global binding. The specific system would thus provide the content that relates to the external world, and the non-specific system would give rise to the temporal conjunction, or the context (on the basis of a more interoceptive context concerned with alertness), that would together generate a single cognitive experience" (Llinas et al., 1998).

The unified experience of consciousness is reduced to coincidence in time at a specific frequency, and the mechanism of this coincidence is a kind of summation. The orchestra must sum up and play a single note together. Nothing is said about different frequencies (melodic and harmonic structures) and patterns of activity in time (rhythms). It all comes down to spike coincidence, simultaneous shooting at the same speed.

At the beginning of the twentieth century, the pioneers of EEG studies noted the transition of instrument readings from one frequency range to another when changing states. Walter Freeman III observed activation in the 40 Hz range in the olfactory system in the 1970s (Freeman, 1978). Then, in the 1980s, the same was observed in the visual system (Gray, Singer, 1989). These observations led to the idea about the leading role of the gamma range in fulfilling the task of temporal coherence, the solution of the binding problem. It has been hypothesized that neurons in different brain regions can encode various stimulus characteristics through synchronized firing at a specific frequency (Singer, 1999). Let's hear from one of the authors of the concept, Wolf Singer, director of the Department of Neurophysiology at the Max Planck Institute for Brain Research.

"Global synchronization phenomena, involving task-dependent coordination of the oscillatory activity of whole cortical areas, have been observed in cats trained to perform a visually triggered motor response. The visual, association, somatosensory, and motor areas involved in the execution of the task synchronized their activity in the frequency range as soon as the animals focused their attention on the relevant stimulus; the strength of synchronization among areas reflected precisely the coupling of these areas by cortico-cortical connections. Immediately after the appearance of the visual stimulus, synchronization increased further, and these coordinated activation patterns were maintained until the task was completed. However, once the reward was available and the animals engaged in consummatory behavior, these coherent patterns collapsed and gave way to low-frequency oscillatory activity that did not exhibit any consistent relations with regard to phase and areal topology. These results suggest that an attention-related process had imposed a coherent temporal pattern on the activity of cortical areas required for the execution of the task ... Another study found recurring phases of heightened discharge synchrony among simultaneously recorded single cells in primate motor cortex that were precisely correlated with the animal's expectancy of a "go" signal. Remarkably, these epochs of enhanced synchronization were not associated with measurable changes in the cells' discharge rates, further evidence that synchrony and discharge rate can be independent ... These findings have recently been extended by the demonstration that during a face recognition task, populations of neurons do not only synchronize locally on the basis of oscillations, but also phase lock across large distances with zero phase lag. A particularly interesting finding was that these large-scale synchronization patterns dissolve and settle into new configurations at exactly the time when subjects had recognized the pattern and prepared the execution of the motor response" (Singer, 1999).

From the TTT point of view, the data indicate that the activity of primary integrators (somatosensory zones), higher integrators, and motor effectors is synchronized during the performance of cognitive and motor acts. There are clear signs that this activity is being prepared based on a projection, which ensures the acceleration and efficiency of differentiation and integration of incoming signals, creating coherent temporal patterns. There are indications that these patterns acquire a stable configuration during training but are subject to active reorganization. We see that the synchronization of parallel channels makes it possible to create convergence at the stage of introjection and transduction in sensory pathways, modulation and primary differentiated integration in somatosensory zones. Likewise, it facilitates the transmission of representations down the network.

The following picture of the relationship of high and low frequencies in the system and their functional role is emerging: the predominance of low frequencies in states of rest and high frequencies coming to the fore with active cognitive and motor activity. However, there is no strict correlation of synch in the gamma frequency range with the binding problem. It is solved by synchronization across the entire range. Moreover, the overall binding is done by low frequencies. That is precisely why they retain their basic pulse in all states, do not disappear anywhere, either with deep dissociation or with vigorous activity. They disappear together with the complete and final disappearance of consciousness in the system (death).

By the way, Singer himself notes that "synchronization is associated with an oscillatory patterning of the discharges, the frequency of these oscillations covering a broad range and exhibiting a marked state dependence" (Ibid). And in the examples cited by him, we can see that, for example, synchronization in the olfactory tract while processing odors occurs at the level of 20 Hz (conventionally attributed to the beta range). But we need to take a closer look at the notion of synchronization in Singer's model. The title of Singer's article is: "Neural synchrony: a versatile code for the definition of relations?" Synchronization turns out to be synchrony of simultaneous firing. Bursts of coinciding spikes somehow encode a signal. It even turns out that there are two codes: the firing rate and its simultaneity. It is striking that both of these phenomena are taken for a code, while neither the speed nor the simultaneity of discrete impulses can be a "versatile code" in principle. Even if the change in rate has a pattern over time, and the same is true for simultaneous firing, it does not mean that this pattern can encode multiple signal parameters.

The musical analogy helps again. Imagine that your ears are tuned only to the perception of changes in the tempo of the music and to the moments when musicians simultaneously play in the same frequency range. You do not hear pauses between notes, nor transitions from one melody to another, nor the contour of even one melody (changing frequency), nor the change in instruments, nor the part of each of them because you have a frequency limit. You do not perceive any rhythmic or melodic pattern or the overlapping of rhythms and melodies (rhythmic and harmonic depth). You hear bursts in the same range and perceive the speed of these discrete bursts. It's hard to imagine but try. It did not work out? It is no surprise. The music code does not work this way. There is simply no such code. If you imagined something, then it has nothing to do with music. Now imagine that you are a scientist who observes the brain's activity and sees nothing but the tempo and the simultaneity of firing in a particular range. In the first case, you remain deaf to the music and its meanings; in the second, you stay blind to the brain's code.

Imagine that you decided to analyze the music of different genres and identify patterns. If you take one parameter — tempo, you get a certain picture, and you can even interpret it. It will reflect part of the information: some genres can be faster on average, and others slower. But will this be the code of the genre, its message?

If we analyze the frequency characteristics, then the picture will still be average and approximate but giving broader information. Thus, rock music will stand out for its midrange boost: for all the importance of the bass part, guitars, keyboards and vocals are the main frequency spectrum of this music. The techno style will stand out in the low range — the bass drum and the bass guitar. In hip-hop, it will be the same as in rock: a selection of midrange, but with more pronounced lows, creating the central groove. Classical music will fall off on the highs since these are live instruments, and there will be a lot of mid and low tones. In folk music, with its emphasis on the voice and live instruments, the highs and lows will be very delicate, and the middle will be the basis.

Suppose we increase the degree of resolution of our analysis and introduce the frequency parameters of the instruments. In that case, we will get an even clearer picture: which instruments are present and which ones prevail. Furthermore, if we begin to analyze the melodic and rhythmic patterns, we will understand which structures prevail in this or that music, where there are intersections between genres, how they "talk" with each other, borrow from each other, learn. We can even highlight the basic common patterns. Messages in unknown languages are deciphered in exactly this way: determining repetitive patterns that may be the key to the rest of the code. This is how the Egyptian hieroglyphs were deciphered; this is how the allies broke the code of the German encryption machine "Enigma" during WWII.

It will also be an analysis of averages, but how much closer to the actual message than the average speed. We will begin to understand something, even if we are deaf, have never heard music, and its code is an enigma for us. We can then take a separate piece and analyze it in even more detail and identify frequency and phase characteristics, melodies, harmonies, and rhythms. We will see how one song differs from another, even if we have never heard them. But it is even better if we look at the result of the analysis and listen to it. Literally: not only look at the monitor screens of devices that analyze brain activity but also listen to these signals.

This is what our brain does when it perceives music and other signals. It analyzes the environment code for all possible parameters — frequency, phase, tempo, amplitude. So maybe we should learn from our own brain and decipher its code in this way? Throughout the history of philosophy, psychology and the human sciences in general, many authors wrote that this is an insoluble task: consciousness cannot cognize itself, the subject cannot become an object. Or even this way: the brain is too complex to know itself. But there is a good saying: never say never. Everything has its time and place.

Fixation on the tempo parameter is a dead-end direction of research, as well as a fixation on synchronization as a simultaneous firing. As elements participating in the wave process of energy-information transmission, oscillators must come to a common "denominator" for synchronization to take place. But they do not have to fire together. A reasonably stable phase and frequency relationship within the synchronization region is enough.

Let us recall the classic example of synchronization of oscillators, from which the study of the physics of this phenomenon began. Huygens observed the coupling of two pendulums in antiphase and called it sympathy, not antipathy. He was intuitively correct: regardless of whether the oscillators are in phase (zero phase difference), in antiphase or in another stable phase relationship, they sympathize-synchronize. The key feature is the stability of the phase coupling.

Now imagine a neuroscientist who observes the activity of neurons and treats synchronization as a firing coincidence. Those that fire together are synchronized for him; those that do their business somehow differently are unsynchronized. That's the whole concept. But this is not a concept but an impasse because the hypotheses emanating from it contradict the physics of the synch process.

And a technological problem appears. MEG and EEG record the total activity of the population since it creates waves of sufficient power to be registered by the device. The antiphase or other variants of the phase difference will either be past the recorded power level or will not represent any coherent picture on the monitor screen, although the nuances of phase coupling for the system itself are the basis of being. When analyzing one frequency level, a considerable part of the information is not reflected. The frequency coupling of different levels of the entire range, the harmonic depth of the music of the Mind are generally not included in the analysis. And again, it is necessary to emphasize: the point is in the concept, and technologies will come sooner or later.

There is one more significant problem with the hypothesis about solving the binding problem as simultaneous firing in a narrow frequency range. If we imagine that such synchrony can create a universal code and combine streams, then the issue of differentiation remains. How can simultaneous shots create a model of reality and even a single representation, which would combine all the parameters of the signals while keeping each parameter and each signal its unique place "on the shelves" of consciousness?

Singer proposes a solution to the problem: "We shall address this grouping strategy as "binding by convergence" or "binding by conjunction cells." This coding principle is also known as "labeled line coding" because the responses of a given unit have a fixed label attached to them; they always signal the same conjunction of input signals. The complementary strategy for response binding relies on dynamic selection and grouping of responses. Here, responses are bound by jointly enhancing their saliency relative to other, nonbound responses. Enhanced responses have a stronger impact on downstream processes than nonenhanced responses and therefore dominate subsequent computations. Thus, the results of these computations will reflect the specific configuration of features to which cells with enhanced responses are tuned. We shall address this selection and grouping strategy as "dynamic binding" and the associated coding principle as "relational coding" or "assembly coding," because here the information about a particular conjunction is contained in the dynamically adjustable configuration of the enhanced responses of distributed neurons" (Ibid).

He dwells on the mechanism of phase regulation of the neuron membrane potential oscillations by fine-tuning of NMDA receptors. These are indeed essential phase regulators. We have discussed their function in previous parts of the study and will be returning to them later in connection with their role in pathologies (more on that in "Part Eight. Dissonances of the Mind"). Here one question is important for us: is all this fine-tuning, phase modulation intended only so that neurons can fire simultaneously in a narrow frequency range? Why not? It is a working hypothesis. Moreover, the empirical evidence seems to be confirming it. Neurons can generate simultaneous spikes within milliseconds, creating general activity in a given frequency range almost instantly. Singer described empirical evidence of such a process as proof of his theory that the unity of consciousness is based on simultaneous shooting in the gamma range (even more narrowly — at a frequency of 40 Hz).

It is impossible to deny what is seen from the EEG and MEG measurements. The devices register simultaneous activity formed by the currents of sufficient power. Suppose this activity is what creates representations (Singer calls it content). How it does the job remains behind the scenes of the concept. Singer's model has no place for coding issues at all, even though he calls synchrony a "versatile code." Joint work can create stable connections (Hebb's rule "fire together, wire together"), but what does the code have to do with it? Code is not a connection. Connection is just a means of transmitting the code. Singer does not address the most crucial question: what are representations physically? Accordingly, he ignores the following questions: how are they formed, what is the mechanism of signal transduction and coding?

In addition to the question of the code, there is also the problem of the speed of its transmission along the entire chain from integrators to effectors and vice versa. Let's take visual modality as an example. Studies have shown that the speed of response to stimulus in the associative zones of the cortex is in the range of about 0.1 sec (Thorpe, 1990). Now let's calculate the number of synaptic connections on the way from the retina to the high integrators of the temporal lobe: two in the retina (transducers), one in the lateral geniculate body of the thalamus (modulator), two in each zone V1, V2 and V3 of the visual cortex (primary integrators) and one in the lower temporal lobe (intermediate integrator). A signal has to go through 10 synapses in 0.1 of a second. That makes 0.01 of a second for each. If we proceed from the hypothesis that the basis of consciousness is 40 Hz, then this is forty spikes per second. Simple arithmetic shows that at this frequency they have no time to do anything, even give one shot, not to mention the formation of a representation with all the nuances of parameters. Even at a higher frequency, neurons have only one or two spikes at their disposal. So, there is no time for transmission and no time for generating a code of average firing rate. From whatever side we approach, it does not work. But the system works. Hence, there is something wrong with the model.

Singer wrote: "Cortical networks should be able to operate with the required temporal resolution, because otherwise they would not be able to maintain synchrony at 40 Hz to begin with" (Ibid). Neurons can work at this frequency and even higher. But if each element in the chain produces a spike with no information in its structure, then even if we forget about the entire signal processing chain and assume that all neurons fired synchronously, the same questions remain. How in a split second the brain manages to form a full-fledged representation? How is information density combined with instant transmission? If the number of participants creates the code's density, how does all this information not mix into one mess, given the convergence of channels?

Singer described it this way: "In addition to their flexibility and virtually inexhaustible coding capacity, dynamically bound population codes have the further advantage that computational results can be represented at all stages of processing by population codes that have the same format. The only changes concern the nature and complexity of the feature conjunctions that are represented explicitly by the neurons forming the respective populations. Thus, distributed representations can be mapped directly onto other distributed representations, which should facilitate polymodal integration and sensory–motor coordination, thus circumventing bottleneck problems" (Ibid).

It is just a description of the fact of life: the brain really has a flexible and capacious code of the same format throughout the entire system; distributed representations do overlap, creating an integrated polymodal worldview. The paradox is that Singer's model describes a code that has no capacity or flexibility and integration mechanism as the synchronous firing of in-phase shots with the same rate, which clearly creates a bottleneck problem. The model does not correspond to the physical and technological reality of the brain.

Instead of a complex but effective and dynamic mechanism for synchronizing oscillations with different frequencies and phases, and the interaction of the waves of these oscillations, such models speak of the coincidence in time of discrete identical spikes. But neurons do not have to fire together for synchronization to occur, and the waves consist not only and not so much of oscillators in one phase.

There is another paradox. Neurons are not as fast as average firing rate theorists want them to be, but fast enough to control the body at a temporal resolution of tenths of a second or even quicker. But at the same time, they can do it at vibration frequencies even lower than 40 Hz. Indeed, in the range around this frequency, many processes occur, but a vast number also go on at lower frequencies. Moreover, the basic sync pulse is an order of magnitude slower.

There is only one way out of both paradoxes: changing the code model and the integration model. Suppose we proceed from the Symphonic Neural Code hypothesis. In that case, the "shot" of a neuron becomes a note and contains a sufficiently large amount of information so that a complex representation could be encoded in the ensemble's play, even within one short measure. And suppose we proceed from the hypothesis of synchronization as coupling of oscillations with various frequency and phase ratios. In that case, we find an answer to how representations map onto each other, and neurons do not have to fight hard to "squeeze in" with their shot in a narrow bottleneck of 40 Hz.

Singer wrote: "In conclusion, I believe that the theoretical implications of the synchronization hypothesis and the data available to date are of sufficient interest to motivate further examination. The application of the new methods required to test the hypothesis will undoubtedly provide new insights into the dynamics of neuronal interactions. If it then turns out that the hypothesis falls short of the real complexity — which is bound to be the case — we will have learned something about the role of time in neuronal processing that we would not have learned otherwise" (Ibid).

This hypothesis really falls short of the true complexity of the binding problem and does not reflect the reality of brain functioning. Moreover, empirical studies have shown that there is no direct correlation between neural synchrony as simultaneous firing and perceptual binding (Thiele, Stoner, 2003; Dong et al., 2008). Any model is bound to be short of full description and cannot cover everything once and for all. The range of the model is not a problem if it reflects the actual physics of the described process. The problem begins when the model contradicts it. We have to differentiate between a model that falls short of the complexity and a model that leads to an impasse. We should stop trying to go deeper and deeper into the dead-end but look for other paths.

The authors of a review article called "Synchrony Unbound" summed up the problems: "The theory is incomplete in that it describes the signature of binding without detailing how binding is computed. Moreover, while the theory is proposed for early stages of cortical processing, both neurological evidence and the perceptual facts of binding suggest that it must be a high-level computation ... Nonetheless, the theory has sparked renewed interest in the problem of binding and has provoked a great deal of important research. It has also highlighted the crucial question of neural timing and the role of time in nervous system function. The problems that gave rise to the theory are still important problems that remain to be solved, and it is certain that the efforts of the theory's proponents and opponents will advance our knowledge both of higher visual functions and of the algorithms used by that most enigmatic of computers, the cerebral cortex" (Shadlen, Movshon, 1999).

It is true that previously most of the theories were concerned only with the spatial aspects of the brain's functional-anatomic structure. But space and time are conjugate variables, and we cannot ignore one or the other. The attempts to explain how the binding problem is solved by the architecture of neural pathways or by coincidence in time (synchrony) are bound to be incomplete. They will also fail to be up to the task if they ignore the algorithms and computational processes that create the information flow that needs to be bound.

We should not miss any part of the full technological chain: information encoding, transmission, storage, retrieval, comparison of the incoming perceptual data and accumulated representations, and update of the reality model. This is a constantly flowing iterative algorithm, and all its stages happen in the same operational space and time of the brain. This technology is founded on the physical mechanisms that manifest themselves in physiological processes. Computational and physical phenomenology cannot contradict each other as the brain is a physical device that encodes the signals of the world and combines encoded representations into a model of reality that has to be unified and differentiated at the same time. Any model of the brain has to deal with all these aspects.

Let's go back to physics and recall that one of the main postulates of the classical wave theory describing the propagation of waves in a medium is the Huygens-Fresnel principle, which is formulated as follows: each element of the wavefront can be considered as the center of a secondary disturbance that generates secondary waves, and the interaction of these waves will determine the resulting wave structure at each space point. The principle is based on the idea that a wave can have one source of origin, but as it moves, each element of the propagation medium becomes a new source of wave motion.

It was initially introduced by Christian Huygens in 1678 when studying light waves. Augustin Jean Fresnel, in 1815 supplemented Huygens' principle by introducing the concept of coherence and interference of waves. Coherence (Latin cohaerens — being in communication) is the correlation of several oscillatory processes. Oscillations are coherent if their phase difference is constant over time. It means that at different spatial points, the fluctuations occur synchronously, i.e., the phase difference between two points does not change (phase coupling).

Often this principle is reduced to the fact that the superposition of waves creates an interference pattern as a mutual increase (constructive interference) and a decrease (destructive) of the resulting amplitude. The interaction of waves is narrowed down to two options. Stable (stationary) interference is formed if the waves have the same frequency (unison) and the phase difference is constant. If the phases coincide, then the maximum increase in amplitude occurs; if waves are in antiphase, they cancel each other out. With unstable (non-stationary) interference, all options are in the interval between the two above extremes, and there is no constant phase relationship.

In this study, we consider the processes of wave interaction in a broader sense, and the term "interference pattern" should be extended to "synchronization pattern" since we are talking about all interaction options with frequency coupling at any synchronization order, and not only at 1:1 unison.

Although Huygens is considered the discoverer of synchronization, neither at the time of the formulation of the principle in the 17th century nor during its development by Fresnel in the 19th century, there was no talk about the mechanism of interaction. But now, it becomes evident that the principle describes how the universal mechanism of synchronization manifests itself: all the phenomena of wave propagation result from the synchronization of oscillators. As the primary source of the wave, the initial oscillation interacts with subsequent oscillators in the propagation medium. The correlation of frequencies and phases creates a stable synchronization pattern that simultaneously preserves the parameters of the source and includes the parameters of all elements involved in the propagation.

The Teleological Transduction Theory expands the Huygens-Fresnel principle to the brain. It also develops it further by adding to the mechanism of wave propagation, their coordination mechanism: synchronization as coupling of various frequency modes.

Hypothesis:

Each act of the Mind, from basic sensory-motor to higher cognitive representations, is physically a propagating wave in which neural network elements tuned to its frequency and phase parameters take part as wave sources and as centers of secondary perturbation. The resulting wave pattern is determined by the superposition of all participating oscillations achieved via synchronization as cross-frequency coupling based on the harmonic frequency ratios. Such a universal physical mechanism makes it possible to perform a huge number of operations with great speed and efficiency, in parallel and sequentially, creating a unified model of external and internal reality while maintaining the differentiation of each representation. We can say that the brain binds our world picture and the picture of us in this world by harmonizing itself.

This *Binding-by-Harmony Hypothesis* elegantly resolves the issue of differentiation and cross-modal integration of the reality model and the unity of Personality which are created by a normally functioning brain. From this follows a reverse assumption that the disruptions of sync are behind dissonances of the Mind that we call mental pathologies. Harmony is health, dissonance is an illness. This is a physical, not a metaphorical description. We need this physical bridge to cover the explanatory gap between mental and physiological phenomena. Otherwise, our model of the normally functioning Mind and its pathologies will lack explanatory power. Without such power any model is useless.

The authors of one review article wrote: "Currently, neuropsychiatric disorders are classified at a phenomenological level, and most biomarkers of these disorders do not capture the heart of the underlying network dysfunction. The spectral fingerprints of neuronal interactions may turn out to be sensitive markers of such disorders. Indeed, a rapidly growing body of evidence indicates that local oscillations and their large-scale coherence are altered in various diseases" (Siegel, Donner, Engel, 2012).

What the authors call "spectral fingerprints" are harmonies of the Mind, and the interaction of these "fingerprints" is synchronization. The authors quite accurately point out that the modern approach to the Mind's pathologies is phenomenological and describes their manifestations without penetrating fundamental mechanisms. We will deal with this issue in detail in further parts of the study.

CHAPTER 6

LOOKING FOR HARMONY IN THE BRAIN

How do all those neurons simultaneously get together in a virtual instant, and switch from one harmonious pattern to another in an orderly dance, like the shuttle of lights on the "magic loom"?

Walter Freeman III

We have been studying brain frequencies for more than a hundred years since Richard Caton in the 19th century used a galvanometer to observe electrical impulses on its surface. Adolf Beck, Vladimir Pravdich-Neminsky and Hans Berger followed suit at the beginning of the 20th century. After their discoveries, studies of various frequency bands and their relation to the state of the brain became one of the mainstreams in neurophysiology. The accumulated data is enormous. But do we have an understanding of how all these frequencies relate to each other and to the processes of the Mind?

Here is the opinion of the authors of a review article: "Major unifying hypotheses regarding the physiological origins and functional relevance of the different frequency ranges are still lacking, although suggestions have been made … Although all these hypotheses seem highly interesting and partly supported by experimental evidence, we do not yet have a coherent picture on why the brain oscillates at such a diverse range of frequencies and how these different oscillatory processes functionally complement each other … Resolving these issues would also require a better understanding on how oscillatory processes with different carrier frequencies … might actually interact" (Engel, Fries, 2010).

The key point in the above conclusion is the following: accumulated physiological facts have to be connected to the psychological phenomena by the physical bridge that explains the mechanism. We have looked at many unifying hypotheses within TTT in the previous parts and in this part of the study. We will continue to do so in an attempt to create a coherent picture of the brain. Here we will consider the question of the interaction of different carrier frequencies. Let's start with the terminology issue that has an important consequence for the coherence of the model.

Many neuroscientists call the frequency parameters of the nervous system "brain rhythms." From the TTT perspective that uses the music analogy, frequency patterns and their interactions are harmonies but not rhythms. Strictly speaking, rhythms are durations and sequences of notes. We will deal with the notes of the brain in further chapters. For now, we will concern ourselves with the harmonies of the brain. Let's see if this change of terminology that may seem subtle will lead us out of the currently observed theoretical impasse.

As an example of such an impasse, we will take a book by György Buzsáki called accordingly "Rhythms of the Brain" (Buzsáki, 2006). He wrote: "While we know quite a bit about neurons, the building blocks of the brain, and have extensive knowledge about their connectivity, we still know very little how the modules and systems of modules work together. This is where oscillations offer their invaluable services ... The same mechanism giving rise to different frequency bands ... ought to be referred to by the same name, even though the dynamics underlying the rhythms may be different. Unfortunately, the exact mechanisms of most brain oscillations are not known. As an alternative approach, ... there might be some definable relationship among the various brain oscillators" (Ibid).

This is where the central problem arises. A lack of understanding does not mean that the mechanism can be bypassed when trying to build a model of brain oscillations. Otherwise, the model will turn into a "castle in the air" that has no support in reality. The assumption about the mechanism must be tested empirically and corrected. Only in this way does ignorance become knowledge. We may not understand the mechanism initially, but we must build hypotheses about it and test them so that misunderstanding turns into understanding.

Buzsáki's alternative approach is to try to find patterns in the data without a hypothesis about the underlying mechanism. This approach is often taken in research in the hope that an accidental discovery of the truth will occur. The problem is that such logic leads to the situation that even if there is a discovery, it is impossible to say what exactly was discovered. It's like looking for a black cat in a dark room with no idea what a cat is. Even if you catch something, you cannot be sure if you have found a cat.

The researchers rely on empirical facts and think that it means that their model is correct. However, this is an illusion, because any pattern can be found in the data. But without relying on the idea of a mechanism that could lead to such a pattern, there is a risk of inadvertently manipulating the data to bring them to at least some pattern. Let's see if the researcher managed to get around the risk in this case.

Buzsáki asserts that he has found a pattern in the frequency ranges of the brain: "There is a definable relationship among all brain oscillators: a geometrical progression of mean frequencies from band to band with a roughly constant ratio of e, 2.17 — the base for the natural (Napierian) logarithm. Since e is an irrational number, the phase of coupled oscillators of the various bands will vary on each cycle forever, resulting in a nonrepeating, quasi-periodic or weakly chaotic pattern" (Ibid).

The author is trying to find a physical justification for such a chaotic pattern: "If the center frequencies of the various oscillators had integer steps, the various bands would be vulnerable to unwanted interference" (Penttonen, Buzsáki, 2003).

Indeed, the interference of waves can cause both the termination of the oscillatory activity of the system and the destruction of the system as a result of a sharp increase in the intensity of oscillations. But such a simple interference pattern applies only to waves of the same frequency. In a complex system with a wide frequency range, the interference pattern is also complex.

It does not mean that unwanted interference cannot occur when the frequencies of different parts of the brain coincide. Moreover, such a picture is observed in some pathologies (for example, epilepsy). It is exactly to prevent such phenomena, that a multi-tiered frequency organization of the brain and the presence of positive-negative feedback mechanisms for regulating undesirable effects during the interaction of waves exist. As we showed in the previous volume of research, the brain has to apply various technological solutions to take advantage of the waves as a means of communication and compensate for the disadvantages.

But the proposed scale means that there is no communication as the relationship between the frequency bands turns out to be irrational. Thus, the phases of the oscillations cannot lock, since they change on each cycle in an infinitely nonrepeating way (the number e is an irrational infinite number 2.718281...). According to the model the brain frequency bands cannot synch. But in reality, they do and thus provide communication.

The author confirms this: "Oscillators of different frequencies and different sites can be phase-coupled, providing mechanisms to bring about precise discharges of neurons without direct anatomical connections" (Ibid). Moreover, he calls synchronization the vital mechanism for the brain: "At the physiological level, oscillators do a great service for the brain: they coordinate or "synchronize" various operations within and across neuronal networks ... Coupled oscillators perform the job of synchronization virtually effortlessly. This feature is built into their nature. In fact, oscillators do not do much else ... Yet, take away these features, and our brains will no longer work" (Buzsáki, 2006).

On the other hand, the author wrote: "A critical aspect of brain oscillators is that the mean frequencies of the neighboring oscillatory families are not integers of each other. Thus, adjacent bands cannot simply lock-step because a prerequisite for stable temporal locking is phase synchronization" (Ibid).

How can oscillators perform the job of synchronization that makes our brain work if they cannot phase lock? Buzsáki says that they should be in synch but insists on the frequency scale that cannot synch. What way out of this paradox does the author suggest?

"The 2.17 ratio between adjacent oscillators can give rise only to transient or metastable dynamics, a state of perpetual fluctuation between unstable and transient phase synchrony, as long as the individual oscillators can maintain their independence and do not succumb to the duty cycle influence of a strong oscillator. In the parlance of nonlinear dynamics, the oscillators are not locked together by a fixed point or attractor (phase), but they attract and repel each other according to a chaotic program and never settle to a stable attractor. A main reason for this recklessness is the presence of multiple oscillators that perpetually engage and disengage each other … Despite the chaotic dynamics of the transient coupling of the oscillators at multiple spatial scales, a unified system with multiple time scales emerges … One may speculate that these interference dynamics are the essence of the global temporal organization of the cortex" (Buzsáki, 2006).

Instead of clarity, things get even more confusing. Is interference unwanted, or is it the essence of the organization? Is there coupling or not? On the one hand, the oscillators are not coupled and have no common attractor in their phase space. On the other hand, in this completely reckless independence and lack of interaction, a connection between different levels does arise in some mysterious way.

Interestingly, there is even a mathematical confusion in the above description. In the original article, Buzsáki uses the number $e=2,71$, while in the book, it appears as 2.17. Even if it is a typo, it is very revealing. The calculations of the scale were arbitrary and did not rely on any physical mechanism.

"The frequencies and ranges were then calculated by starting from 50 Hz gamma center frequency and going to slower and higher frequencies separately, since this approach gave the best correspondence to the regression estimations … The constant intervals between oscillation bands on the logarithmic scale (arithmetic progression) correspond to a constant ratio between the bands on the linear scale (geometric progression)" (Penttonen, Buzsáki, 2003).

The choice of the center frequency is entirely arbitrary so that the scale corresponds to the pattern the authors want to see in the data. But the problem is not even the arbitrariness of the numbers but that the scale begins to contradict the physics of the process. To overcome such a discrepancy, the author calls for the "recklessness" of chaos, which is responsible for everything, and creates everything "transiently" in an unclear way.

All these internal contradictions of the model and its inconsistency with reality are the price for searching for a pattern in the data without a physically based hypothesis about what mechanism may lead to the resulting data. This leads to the illusion of empirical validity, while the discovered pattern turns out to be just a manipulation of the numbers. The illusion is so strong that for many years the author insists on the model, which has no physical meaning and contradicts his own descriptions of the functioning of the brain. Here are some quotes from more recent works.

"Neuronal networks in the mammalian forebrain support several oscillatory bands (families of oscillations) that span from approximately 0.05 Hz to 500 Hz. Importantly, there are a number of boundary lines drawn to delineate cortical oscillations which have been empirically found to act relatively independently. The frequencies occupied by these bands have relatively constant relationships to each other on a natural logarithmic scale" (Buzsáki, Watson, 2012).

"Unfortunately, the taxonomy of brain oscillations is poorly developed, and existing terms typically refer to the frequency band that the rhythm occupies rather than its mechanism ... Cross-frequency coupling across the various rhythms, which have a typically noninteger, irrational relationship with each other creates an oscillatory interference, and this interaction is most likely responsible for the brain's perpetually changing activity patterns" (Buzsáki, Logothetis, Singer 2013).

While criticizing the situation in neuroscience as a whole, the authors describe their own approach: they consider frequency bands arbitrarily and in isolation from the physical mechanism. Hence all the confusion. On the one hand, the frequency scale of the brain is irrational and the frequencies of the "neighboring oscillatory families are not integers of each other" and "cannot simply lock-step." On the other hand, "oscillators of different frequencies and different sites can be phase-coupled." On the one hand, the non-integer ratio prevents interference. On the other hand, it creates interference.

Such a description is chaotic. Does this mean that the frequency structure of the brain is as chaotic as the proposed model? It is no coincidence that Buzsáki refers to Walter Freeman's ideas about the chaotic organization of wave packets and Karl Friston's ideas about "stable incoherence." They are about the old theme of the emergence of order from chaos without any mechanism for creating this order. But there are doubts: "It is not clear, though, how stable incoherence (high entropy) can be maintained in an interconnected system, e.g., the brain" (Buzsáki, 2006). Indeed, it is not clear how incoherence can create a coherent system. The question remains unanswered except for general words about complex dynamic instability.

Of course, the interaction of oscillatory processes in the brain has a dynamic character with many transitional states. But that doesn't mean it's chaotic. An attempt to create a model of an interconnected brain system based on the irrational number e can be described in the words of the author himself: a chaotic program without a stable attractor and the recklessness of multiple oscillators. The paradox is that the author describes the pathology of decay, although he claims that he is creating a model of an interconnected system. Not surprisingly, the model does not correspond to the reality of the brain functioning which is not chaotic and not reckless in a normal mode of operation.

The authors of one study took the EEG spectroscopy data from 1424 healthy subjects and used a novel method to determine peak parameters in the frequency range from theta to beta and investigate their interrelationships. They reported: "Peaks in the theta range occurred on average near half the alpha peak frequency, while peaks in the beta range tended to occur near twice and three times the alpha peak frequency on an individual-subject basis. Moreover, for the majority of subjects, alpha peak frequencies were significantly positively correlated with frequencies of peaks in the theta and low and high beta ranges ... We have shown clear interdependences between the frequencies and amplitudes of peaks in different bands in this condition, frequencies of many peaks following an approximately harmonic progression ... It is not consistent with any of the following proposals: a geometric progression with a peak spacing of Euler's number or the golden ratio ... More generally, to our knowledge there is no model of purely cortical oscillations that predicts the observed peak relationships" (Van Albada, Robinson, 2013).

Let's try again to use the musical analogy to create a model of brain oscillations that would predict and explain the observed harmonic relationships. It may happen that the harmony of the model will reflect the harmony of the normally working brain.

Buzsáki is aware that cross-frequency coupling means integer ratios: "Cross-frequency interaction is referred to as cross-frequency phase-phase or n:m coupling, when there is an integer relationship between the frequencies of the two rhythms ...The phenomenon of cross-frequency coupling firmly demonstrates the hierarchical organization of multiple brain rhythms in both space and time and implies that time in the brain is represented at multiple correlated scales" (Buzsáki, Watson, 2012).

This description seems to be logical and physically grounded. If not for a subtle nuance that ruins all logic: frequencies are called rhythms. It is a category error (logical fallacy). Frequency is a space domain parameter. Rhythm is a time domain. They are conjugate parameters that are combined in any continuous signal but they should not be mixed up when we analyze this signal. For an intuitive explanation, let's take the musical staff as an example. All notes are put on the staff along the horizontal axis representing their development in time as durations and sequences. This is the rhythm. But they also have pitches that are represented by putting the notes on the ladder of the vertical axis representing space parameter. We can also think of the staff as a spectrogram where the frequencies are reflected on the space axis and phases on the time axis.

So, when the author writes that cross-frequency coupling is the hierarchical organization of rhythms in both space and time, it is a mix-up of different categories that leads to a chaotic model. He writes that time in the brain is represented at multiple correlated scales. But these scales are not frequency scales. They are rhythmic structures that the author does not even touch upon. His research is fully devoted to the frequencies of the brain, but not rhythms as durations and sequences of the notes of the music of the mind. This picture is characteristic of the mainstream of neuroscience. We have touched upon this problem in the previous parts of the study and will be dealing with the notes of the Mind in further chapters.

Buzsáki also uses musical comparison to describe the interaction of brain elements: "Syn (meaning same) and chronos (meaning time) together make sure that everyone is up to the job and no one is left behind, the way the conductor creates temporal order among the large number of instruments in an orchestra" (Buzsáki, 2006).

Yes, synchrony initially means coincidence in time. But as a general concept, synchronization means frequency and phase locking that is developing both in the space and time domains. Thus, when the conductor waves his hands to outline the basic pulse of a symphony so that various rhythmic patterns can be combined, it is only a part of his job that is visible during the performance and a part of what we call music. The task of a conductor is to prepare the orchestra to perform the score so that it sounds like a symphony not like a cacophony. This score consists not only of rhythmic parts but of multiple frequency patterns that are harmoniously combined according to the laws of cross-frequency coupling. We call these patterns melodies and chords. The rhythmic structure that is represented by the conductor's hands forms only the context of the symphony. The frequency structure forms the content.

Many neuroscientists compare the brain to the orchestra but they use this comparison as a metaphor not as a physical analogy. They are not looking into the details of the physical mechanism that produces the result that we call music. It seems that the author of the irrational frequency scale of the brain falls into the same trap. If the brain is an orchestra, how can it play a symphony if its frequency "families" are not in harmonic relationships? According to his model, the brain frequencies are widely out of tune and some conductor is trying to create temporal order in their chaotic and reckless dynamics. A musical orchestra playing such a cacophony would be dispersed. The conductor will not save the day by waving his hands to denote the beats if the notes played by the musicians are out of tune with each other.

In musical tuning, one can deviate from pure intervals, but within a very small discrepancy, measured in cents. The octave is divided into 12 parts, and each of the resulting semitones is divided into 100 cents. A 1 cent change in frequency is approximately equal to a 0.058% change. Differences in units of cents are acceptable. If the difference is tens of cents, then there is dissonance ("wolf interval"). For example, a wolf fifth differs from a perfect fifth by about 21-24 cents (1-1.5%). Behind this mathematics is the physics of the process. A strong deviation from the harmonic interval means that the frequency ratio is out of the synchronization region. Note that the frequency scale based on the irrational Euler number $e=2.71$ differs from the pure octave interval by hundreds of cents. Using musical terminology, we can say that such a series would be a "wolf octave" of dissonances.

All tuning systems were aimed at creating the maximum range with the minimum number of dissonant frequency combinations. Long practical and theoretical searches have led to the now-standard equal temperament tuning, where each octave is divided into mathematically equal intervals of 12 semitones. This allows to easily shift the pitch of all notes to the desired interval and move from one key to another (transposition and modulation). The listener perceives such a transition as a smooth one, and the sound is almost equivalent to the original key. With minor deviations from pure intervals, such a tuning system simplified the combination of many tones and gave a considerable impetus to the development of harmonic depth.

Physicist and mathematician, Yakov Perelman, wrote: "Musicians are rarely fond of mathematics; most of them, having a sense of respect for this science, prefer to stay away from it. Meanwhile, musicians … come into contact with mathematics much more often than they themselves suspect, and, moreover, with such terrible things as logarithms. Let me cite on this occasion an excerpt from an article by our late physicist Professor A. Eichenwald: "My high school friend liked to play the piano, but did not like mathematics. He even spoke with a tinge of disdain that music and mathematics had nothing to do with each other. "True, Pythagoras found some kind of relationship between sound vibrations, but it was the Pythagorean scale that turned out to be inapplicable for our music." Imagine how unpleasantly my friend was amazed when I proved to him that, when playing the keys of a modern grand piano, he actually plays on logarithms" (Perelman, 1967).

He made a simple calculation. The note C of the "zero octave" is n vibrations per second, the next one will be 2n vibrations, and the m-th octave is $n \times 2^m$ vibrations, etc. If we designate all notes of the chromatic scale with the number p, taking the fundamental tone of C of each octave as zero, then the note G will be the 7th, A will be the 9th, etc., the 12th tone will be C again, but an octave higher. Then, in equal temperament scale, where each subsequent tone has $\sqrt[12]{2}$ times the frequency, each note can be expressed by the formula:

$$N_{pm} = n \cdot 2^m \left(\sqrt[12]{2}\right)^p$$

If we take the logarithm of this formula, then lg Npm = lg n + m lg 2 + p lg 2/12 or lg Npm = lg n + (m + p/12) lg 2. Taking the number of fluctuations of the lowest C as one (n = 1) and transferring all logarithms to base 2 (or simply taking lg 2 = 1), we get lg Npm = m + p/12. From this we see that the numbers on the piano keys are the logarithms of the vibration numbers of the corresponding sounds multiplied by 12.

The frequency of any note can be calculated by multiplying/dividing or adding/subtracting the frequencies of other notes, and the logarithm is just a mathematical operation, where laborious multiplication is replaced by a simpler addition, and division is replaced by subtraction. For example, in 5-limit tuning, the base note C has a frequency of 256 Hz (in equal temperament tuning — 261.6 Hz). We can calculate the E frequency in different ways:

$$f_E = 5^1 \cdot 3^0 \cdot 2^{-2} \cdot f_C = \frac{5}{4} \cdot 256 Гц = 320 Hz$$

or

$$f_E = 5^0 \cdot 3^4 \cdot 2^{-6} \cdot f_C = \frac{81}{64} \cdot 256 Гц = 324 Hz$$

All this algebra is about the physics of energy vibrations and their interaction. Mathematically speaking, the musician plays on logarithms, but physically he produces oscillations with various frequencies. The musical frequency series is based on integer ratios (pure intervals), and the logarithm is binary (base 2), since the series is iteratively closed by an octave (frequency ratio 2:1). The harmony of tones, expressed through ratios of integers, is not a mathematical abstraction, created arbitrarily for the beauty of the logarithmic pattern, but reflects the natural physical mechanism of cross-frequency interaction with clear phase coupling parameters.

Here is a simple illustration of such a process:

$$f_2 = 2 \times f_1$$

f1   f2

Frequencies at 2:1 octave ratio are anti-phase in one peak of the cycle and in-phase in the other. Overall, there is coupling that means a stable phase relationship. This illustration is ideal. In reality, there is a certain freedom of phase shift within the synchronization region and, accordingly, the ratios are not fixed, but only tend to integers. Such a series is not recklessly chaotic, but sensibly harmonious. It creates order in the interaction of various frequencies, which we call notes, and forms an interconnected system of sound vibrations, which we call music.

The TTT offers an alternative approach that shows a way out of the theoretical impasse of an arbitrary classification of brain frequencies. Perhaps this approach will avoid the risk of internal contradictions of the model and its contradictions with reality. Based on the idea of the physical mechanism of the process, we can assume that the frequency series of the brain has an octave periodicity since 2:1 forms a metastable, consonant frequency ratio with a coherent phase portrait, a stable attractor, and a wide synchronization region. The "musicians" of the brain orchestra, as interacting and synchronized oscillators, play on the logarithm and this logarithm has base 2.

With this approach, the description of the interaction of frequencies will be radically different from that proposed by Buzsáki. Let's try to paraphrase a quote from his book. The 2 ratio creates transient and metastable phase-locked dynamics in which individual oscillators can interact while maintaining independence. In the parlance of nonlinear dynamics, the oscillators are not locked together by a fixed point, but in accordance with a harmonic program leading to a stable attractor. Due to the harmonious dynamics of the transient coupling of the oscillators, a unified system with multiple spatial and time scales emerges. One may speculate that these synchronization dynamics are the essence of the global organization of the brain.

Hypothesis:

The nervous system is a set of oscillators with different frequency and amplitude characteristics, determined by the parameters of the elements participating in the oscillatory process. These parameters create a wide frequency range in which various bands can perform different functions, and different parts can operate at multiple frequencies. The variety of frequencies, amplitudes and phases of these oscillations is not an obstacle to unifying all these spatial and temporal levels into a single and coherent system since a universal physical mechanism of synchronization as frequency and phase coupling exists to create a structure from such a variety. The distribution of frequency bands in the brain and their ratios obey the universal laws of the interaction of oscillators. The elements of the system tend to adjust their parameters within the synchronization regions approaching integer frequency ratios. The frequency series of the brain is based on an octave periodicity, in which the base is a frequency ratio of 2:1 with a stable phase portrait and a broad synch region.

Zero octave: 0-4 Hz
First octave: 4-8 Hz
Second octave: 8-16 Hz
Third octave: 16-32 Hz
Fourth octave: 32-64 Hz
Fifth octave: 64-128 Hz etc.

The combination of tones (frequency peaks) within each octave and with tones in other octaves tends to harmonic intervals (integer ratios of frequencies). We can call the frequency series of the brain a temperament tuning system of the music of the Mind. Such scale means that multiple brain frequencies interact and phase-lock while retaining dynamic freedom of parameters within synchronization regions. Violation of the harmonious cross-frequency interaction and phase coupling means the disintegration of the system and the pathology of brain functioning.

The frequency series forms a continuum. Therefore, we do not use more subtle divisions and the end of the first octave is designated as the beginning of the second and so on. The beginning is taken as zero oscillations. However, it would be more accurate to define it as a particular minimum frequency at which the vital activity of the organism is maintained. But this is such a poorly studied range that a conditional zero is sufficient for the current classification.

In the proposed scale, octaves are not divided into tones. Only a general principle is indicated, and this table requires further expansion as empirical data on brain frequencies develop. Knowledge of the physical mechanism allows us to predict that further study of the frequency levels of the entire nervous system will lead to the discovery of bands that will have harmonic ratios within the octave.

Perhaps a musical analogy will help us here again. In the theory and practice of music, tones are organized with the help of simple algorithms. Here is an example of such an algorithm called the "circle of fifths":

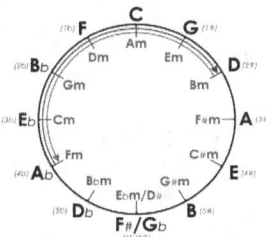

Pitch classes are indicated on the circle: major keys along the outer circle, and minor keys along the inner circle. The "magic" of the circle is that each key can be turned into another by simple mathematical operations precisely because they are all related by harmonic intervals and are symmetrical with the respect to the basic octave interval. Moreover, such an algorithm allows you to do transposition and modulation without any calculations, because all the mathematics is already embedded in the arrangement of keys in a circle. Anyone can translate one key into another simply by turning the circle.

You can take standard sets of harmonic combinations (chords) or sequences (melodies), or you can diversify the structure by borrowing from neighboring tonalities. But in any case, the rigorous mathematics of the circle of fifths will leave all ratios consonant. All popular music is written according to this scheme. A simpler one takes basic combinations; a complex one creates variations. The set of elements is limited, but it has the potential for endless variety. The circle is both an algorithm for creating musical forms and a convenient tool for improvisation. There are even standard improvisational moves. For example, in jazz, tonalities usually change clockwise, and in rock music — counterclockwise. What seems to the uninitiated to be a free and pleasant flow of sounds is a strictly verified structure of consonant combinations and sequences based on the harmonic laws of oscillatory processes.

Considering the astronomical number of oscillators in the brain, the variety of amplitude-frequency characteristics is enormous. But this does not mean that it has no structure and cannot be described by an easy-to-use model. Moreover, if there were no structure in it, then the work of such an orchestra would be impossible. We need a clear concept that will allow bringing all the variety of data into a practically convenient model with explanatory power. Understanding mutual transformations of frequency ratios will help answer long-standing questions: why do specific frequencies interact in various functional states precisely this way and not otherwise; what is the normal structure of the brain frequencies; what combinations lead to pathologies (more on that in "Part Eight. Dissonances of the Mind").

Buzsáki noticed: "An oft-heard marketing slogan these days is that we have learned more about the brain during the past decade that during the previous history of humankind. This may be true regarding the volume of factual knowledge. But discoveries are not just facts. They are ideas that simplify large bags of factual knowledge" (Buzsáki, 2006). His idea of an irrational frequency structure did not simplify but confused and even contradicted the accumulated bags of factual knowledge. For a successful interaction, the frequency ratios should be not chaotic and irrational, but simple and rational. Thus, non-overlapping high-frequency patterns representing various signals can be nested within a low-frequency cycle.

Can we substantiate the hypothesis proposed in the framework of the TTT about the harmonic frequency scale of the brain with the available experimental data?

First, we see that with such a mathematical distribution based on a physical sense, the ranges surprisingly almost coincide with the bands accepted in neuroscience, which were obtained as a result of many years of studying the oscillations of the brain in its different zones and under various states: delta, theta, alpha, beta, low gamma, medium gamma, high gamma. Second, there are numerous studies that confirm harmonic frequency relationships in the brain. Here we will limit ourselves to a very small range of physiological facts that acquire a physical explanation if we accept the hypothesis of a harmonic frequency series.

The authors of one review article took insects' olfactory system as an example. They noted: "Relatively small size and reduced complexity, is ideal for studying the mechanisms that underlie oscillatory communication, and their functional consequences, in detail" (Schnitzler, Gross, 2005). Briefly, the functional-anatomical scheme is as follows: the antennal lobe receives streams from the olfactory receptors, modulates and transmits them to the mushroom body and the lateral horn, which integrate information, create representations of signals and store them. The activity of these areas is phase coupled. "Lateral horn inhibitory interneurons fire about half a period after the phase-locked projection neurons. Consequently, inhibitory postsynaptic potentials (IPSPs) to mushroom body neurons that arise from input from the lateral horn occur half an oscillation cycle after the excitatory postsynaptic potentials (EPSPs) that arise from direct input from the antennal lobe. The strong IPSPs lead to collective inhibition of mushroom body neurons during half the oscillation cycle" (Ibid). We see the play of two push-pull forces of activation and inhibition creating different oscillation frequencies, the combination of various frequencies and phase coupling with stable parameters of integer ratios (2:1 in this example).

The authors of one study created a computer simulation of a simple network of three activating and one inhibitory neuron with characteristics close to those observed in reality. Initially, when the neurons were not combined into an interacting system, they produced spiking with random phases. But after establishing connections their activity showed a coherent synchronized structure. "The study suggests that fast spiking inhibitory neurons mediate synchronization of cortical excitatory neurons, creating large areas of stable synchronized network oscillations where excitatory cells are phase-locked to the field at rational frequencies" (Rulkov, Bazhenov, 2008).

In another study, the authors used a new method for the extraction of cross-frequency synchronized components — Generalized Cross-Frequency Decomposition (GCFD). It works for any f1 and f2 whenever f1/f2 is a rational number. They validated the new method in simulations and tested it with real EEG recordings. The problem with analyzing EEG signals is that they are non-sinusoidal and are a sum of waves at the fundamental and harmonic frequencies. This internal spurious coupling can be a side-effect of a complex wave structure of the EEG signal but not a result of neuronal interactions. To control for this effect, the authors calculated the pattern divergence between the spatial patterns of the reference signal and the synchronized components. If the patterns were similar they considered them to be harmonics. Using alpha range as base frequency they found many patterns with high cross-frequency phase-locking value for 2:1 ratio (Volk et al., 2018).

Another group of researchers identified brain-wide cross-frequency coupling (CFC) networks at mesoscale resolution from stereoelectroencephalography (SEEG) and at macroscale resolution from magnetoencephalography (MEG) data. They developed a novel graph-theoretical method to distinguish genuine CFC from spurious. It showed that genuine interareal CFC of theta and alpha oscillations with higher frequencies in large-scale networks connecting anterior and posterior brain regions is present in both SEEG and MEG data. The strength of coupling was also predictive of cognitive performance in a separate neuropsychological assessment (Siebenhühner et al., 2020).

The evidence that theta oscillations interact with gamma in a certain order was found in studies of hippocampal neurons of rats orienting in space (O'Keefe, Dostrovsky, 1971). Several theta cycles occurred as the rat ran in some area, and its brain encoded representations of that space. The researchers found that with each successive cycle, neurons fired at an increasingly early theta phase. Subsequently, a name was even given to this phenomenon: phase precession (O'Keefe, Recce, 1993).

Spikes of individual neurons can occur in different phases of the cycle. But the distribution of the phases of different neurons and populations is not random and forms a systematic relationship with phases of the theta range. The studies revealed the stable relationship between the spike phase and the rat's position in the maze. It is considered one of the first evidence of the temporal code and the importance of oscillatory parameters.

Subsequent studies have shown that each time a rat finds itself in a completely new place requiring the creation of new representations, the neurons again gradually come into phase coupling with the theta frequency (Bose, Recce, 2001). As the rat moved through the maze, peaks of the oscillations of neurons constantly moved within the theta phase, sometimes getting into it, then leaving it and gradually coming to antiphase, and then falling back into phase. Thus, high frequency adjusted to the slow one. It gave the impression of a constant phase precession, as an advance of the oscillation frequency of these cells in relation to the slow frequency theta. They "ran" ahead of the slow wave phase, but as a result, at the right time, they were either in-phase or in antiphase, which in both cases indicates the achievement of synchronization and phase coupling. Thus, calling phase precession adjustment of phase would be more precise.

The phenomenon of phase adjustment confirms that there is a rhythmic structure of individual spikes that adaptively, dynamically fits into the structure of population activity. These dynamics reflect the process of signal coding and carry information. If we relate the spike time of individual neurons to the theta frequency of the population so that each spike rises in a certain phase from 0° to 360°, depending on when it occurred relative to theta oscillations (0° corresponds to the lowest oscillation phase), we can see that information about movement in the maze is encoded by the spike time in relation to theta rhythm. Such a phenomenon is observed not only when moving through a maze but with entirely different activities, which may not be associated with coding a place in space (Harris et al., 2002).

Sophisticated technologies for recording data from populations of neurons have shown that different cells of the hippocampus are activated in a distinct temporal order during the theta cycle. They called such sequences "sweeps" since the order of activation corresponded to the rat's movement along a trajectory in space.

"Such data show directly that different cells representing different information (i.e., positions) fire at different theta phases. Given that firing is also modulated by gamma, there can be little doubt that a theta-gamma code is used in the hippocampus to represent ordered multi-item messages. An important property of sweeps is that that they are time compressed. Whereas the rat might take ~300 ms to move between positions a and b, cells representing these positions fire ~30 ms apart during a sweep. Furthermore, this time-compressed readout is often predictive, providing a way of rapidly informing downstream networks of the sequence of upcoming places … As the rat runs through a place field, a process that takes several seconds, theta phase becomes progressively earlier (closer to zero phase) on successive theta cycles" (Lisman, Jensen, 2013).

Phase adjustment is found in the brain areas that provide input to the hippocampus and in the areas to which the hippocampus sends its output (Jones and Wilson, 2005; Hafting et al., 2008; Kim et al., 2012; Mizuseki et al., 2009; van der Meer and Redish, 2011). The observed functional connection of the hippocampus with other areas reflects the PAAL algorithm with the projection and introjection flows being compared and the reality model updated. It is no coincidence that phase adjustment was so clearly manifested in the hippocampus as, according to TTT, it is an integrator responsible for new representations production.

But phase adjustment is characteristic not only for spatial orientation and not only for the hippocampus (Malhotra, Cross, Meer, 2012). The change in the phase of the oscillations and adjustment to the low-frequency range was found in monkeys and humans performing various tasks. Interestingly, the phase in which the gamma oscillations occurred differed in the task with and without reward, and the structure of this phase relationship predicted changes in behavior. Coupling between the prefrontal cortex and posterior regions influences working memory, task switching and error correction performance (Brzezicka et al., 2011; Cavanagh et al., 2009; Cohen and Cavanagh, 2011; Palva et al., 2010; Phillips et al., 2014; Liebe et al., 2012: Sarnthein et al., 1998; Sauseng et al., 2005; Schack et al., 2005).

The hippocampal neurons, as well as other brain populations, have various frequency levels. For example, within the gamma range, research and review authors distinguish slow and fast gamma. If we look at the approximate boundaries of these ranges, we will see that they correspond to the predicted ranges of the first and second octaves (32-64 and 64-128 Hz). Experiments with rats showed that slow gamma is associated with memory recall of upcoming trajectories, while fast gamma support encoding of the animal's actual location in real-time (Zheng et al., 2016). Studies using electrode arrays implanted in the brains of rats showed that during sleep, high-frequency bursts of oscillations in the hippocampus corresponded to the same bursts in the parietal, middle, and anterior regions of the cortex, but not in the primary sensory-motor zones. The authors noted that coupling between ripples in the hippocampus and cortex was especially strong during sleep after training (Khodagholy, Gelinas, Buzsáki, 2017).

Overall, the research is consistent with the TTT hypothesis of the place of the hippocampus in the memory formation (see "Part Five. Technologies of the

Mind"). During high-frequency phases of sleep, forming new representations and memory consolidation, as settings of the higher integrating filters impulse responses, continues. The sensory-motor zones (primary integrators) go into a rest mode, and the higher integrators of the cortex and the hippocampus continue their work in a synchronized mode. Oscillations of the low-frequency range play the role of the synchronizing clock. Failures in this basic pulse result in a disruption to the smooth process of forming representations. Pathological changes in memory work when areas directly related to the generation of low-frequency oscillations (for example, the middle septal nucleus) are damaged or exposed to external stimulation confirm this hypothesis.

In the active state, the entire chain begins to work, including the primary sensory-motor zones. Studies with rats show that "in the early stages of the development of a conditioned reflex, theta waves in the hippocampus are ahead of theta waves in the temporal cortex. When the conditioned relation is strengthened, theta waves of the entorhinal cortex are ahead of the waves in the hippocampus ... Stimulation of the septum in rabbits and rats with stimulation parameters that enhance the theta rhythm in the hippocampus accelerated the development of a conditioned reflex and contributed to its preservation" (Simonov, 1981).

From the point of view of TTT, it is PAAL algorithm in action. When the representation forms, the hippocampus waves are ahead of the cortex waves, and when the model is projected, cortex waves are ahead of the hippocampus waves.

Studies have also shown that periods of high-frequency activity during sleep repeat the patterns and correlation of activity in certain zones observed during active actions during the experiment. The authors suggest that "initial storage of event memory occurs through rapid synaptic modification, primarily within the hippocampus. During subsequent slow-wave sleep, synaptic modification within the hippocampus itself is suppressed and the neuronal states encoded within the hippocampus are "played back" as part of a consolidation process by which hippocampal information is gradually transferred to the neocortex" (Wilson, McNaughton, 1994).

During the phase of fast-wave sleep, the brain, remaining in a state of dissociation from the introjection of current signals, repeats the patterns of activity and consolidates the settings of the impulse responses of its filters. This is how new representations are stored in the memory of the system. In sleep, as in the active state, high-frequency activity is modulated and synchronized by a low-frequency one. If there is a stable reference basic pulse that determines the time signature and its beat, then oscillations in other frequency ranges of all participating filters can be placed within this common "grid," creating coherent patterns of synchronized representations as harmonies of the Mind.

The researchers recorded the activity of single neurons and gamma oscillations of populations of neurons in different cortical zones of mice and rats. "Laminar analysis of neocortical gamma bursts revealed multiple gamma oscillators of varying frequency and location, which were spatially confined and synchronized local groups of neurons ... A general feature of cortical oscillations is that slow rhythms engage large areas and effectively modulate the more localized and shorter-lived fast oscillations ... Theta-modulated cells have been found in the entorhinal cortex, perirhinal cortex, cingulate cortex, prefrontal cortex, amygdala, anterior thalamus, mammillary bodies, the supramammillary nucleus, and the subiculum" (Sirota et al., 2008).

Analysis of cross-frequency coupling between theta and gamma oscillations during new rule learning tasks revealed that high and low gamma were coupled with different phases of theta and coupling dynamics at an early stage when the animals still follow the previous rule and at a late stage when the same animals learn the new rule differed. The authors suggested that this reflects the switching of information processing in the hippocampus by the theta phase and that high and low gamma bands play different roles in "retrospective coding" and "prospective coding" (Nakazono et al., 2019).

In TTT terms, it means that different wave patterns at various frequency levels interact in the PAAL algorithm so that projection and introjection flows can be compared and the reality model updated. Frequency and phase coupling allows the flows to be integrated while retaining specific characteristics. For example, if slow and fast gamma oscillations occur in different phases of theta, then this prevents premature mixing of the current encoding with reproduced representations and also creates conditions for comparison and further "packing" of updated patterns into a general model of reality based on a single rhythmic structure of theta that preserves its phase synchronization in different zones.

What do neurons do when they encode new representations? They combine their oscillations into a common wave, the pattern of which should become this representation. When there is a coupling of phases, we can say that the representation is formed, and the pattern has acquired shape. If a new representation is needed, then the process is repeated with the same general regularities but differs in the details of the parameters of the pattern and the elements of the network participating in it. In musical terms: a new melody is superimposed on the existing ones based on the harmonic cross-frequency ratios and is built into the overall rhythmic structure based on a precise phase coupling with the carrier pulse.

Here is how research confirms this theoretical description. In experiments with rats, the cross-frequency coupling significantly increased with learning, and synchronization within the gamma band was reflected in memorization tasks (Montgomery, Buzsáki, 2007). Moreover, coupling strength predicted whether the rat would make the right choice or not (Tort et al., 2009). Modulation of coupling indicated that new locations were saved in memory (Shirvalkar et al., 2010).

Studies in monkeys showed increased coherence in the gamma band and its phase coupling with low frequencies during working memory tasks (Siegel et al., 2009). Single-unit recordings allowed to analyze neuronal activity that corresponded to different memories and showed that they are produced at different gamma phases. Thus, representational high-frequency content is rhythmically separated within the base pulse of low-frequency context. In an associative learning task, theta and alpha/beta synch between the hippocampus and prefrontal cortex carried separate error- and correct-trial information (Brincat, Miller, 2015).

Experiments using high-density electrocorticographic (ECOG) electrode arrays and multivariate pattern analysis have demonstrated that wave patterns for passively perceived color and color recalled from memory differ in terms of spatial configuration and frequency distribution, but retain phase and frequency coupling (Tanigawa et al., 2022).

In experiments with humans, strong theta-gamma coupling in the cortex and in the hippocampus correlated with long-term memorization (Friese et al., 2012) and tasks for recruiting working memory (Canolty et al., 2006, Maris et al., 2011). Synchronization in the gamma range between cortex and hippocampus predicted success in memory formation (Fell et al., 2001). There is evidence that during a task that involved memorizing a sequence of items a peak in gamma power for each successive item shifts along the phase of an underlying theta rhythm during successful, but not unsuccessful, sequence encoding (Heusser et al., 2016). It could be only a correlation of phenomena and not a causal relationship, if not for the fact that disturbing the parameters of the gamma cycle by a genetic or drug effect on the neurons led to memory deficits (Fuchs et al., 2007, Sohal et al., 2009, Robbe et al., 2006). There have been experiments where slow oscillations were artificially amplified by transcranial electrical stimulation of people during sleep. "Selective interference with the key rhythms underlying the replay of spike sequences impairs memory performance, whereas enhancement of the relevant oscillatory patterns improves memory performance" (Watson, Buzsáki, 2015).

So, if synchronization occurred during a normal cognitive process and changes in the parameters of the oscillators led to desynchronization and disruption of functions, we can reasonably conclude that synchronization is the basis for the normal functioning of the system. And vice versa, if synch is the norm, then desynch means a pathology.

The authors of an article called "Normal and pathological oscillatory communication in the brain" wrote: "Investigating the interplay between different frequencies adds another dimension to the already complex identification of spatiotemporal and frequency-specific neuronal networks and require appropriate techniques for dimensionality reduction. The current linear analysis approaches are likely to allow us to see only the tip of the iceberg of the oscillatory neural communication processes" (Schnitzler, Gross, 2005).

We should not be overwhelmed by the complexity that unravels while we penetrate further into the system. If current approaches show us only the tip of the iceberg, we should just dive deeper. But we need a stable ground not to lose ourselves in this ocean of data. We need a model that has clear and relatively simple principles guiding us in our diving.

Chapter 7

Musical Notes of the Mind

The brain state allows us to reconstruct the conscious state, just as musical notes on paper can be transformed by an orchestra into music we can hear.

Julian Barbour

The unified symphony of the Mind consists of separate notes. In the previous parts of the study, we have already considered the algorithms and technologies forming these notes. Let us repeat the general approach to the work of neurons within the framework of the Symphonic Neural Code hypothesis.

The activity of the neuron has the amplitude-frequency characteristics and the phase portrait of each action potential. It allows the brain to significantly expand the capacity of the code, in which information is contained both in the rhythmic structure of spikes and interspike intervals and in the internal structure of the spikes themselves. A sequence of spikes with specific parameters becomes a melody, and simultaneous synchronized combinations of a vast number of notes become chords and harmonies of the Mind.

Thus, the main point is that the spikes are considered a continuous oscillatory process, notes with a particular pitch and duration carrying unique information in the waveform of each spike. This fundamentally distinguishes the model of the Symphonic Neural Code from the prevailing models in neuroscience, which approach spikes as identical "shots."

Even listing the names of researchers who adhere to this view would require a separate volume. We will confine ourselves to one quote from the co-author of Predictive Coding Theory (Rao, Ballard, 1999). This is how Rajesh Rao describes neuron spike in his lectures on computational neuroscience: "There is a very stereotypical shape of the action potential, which is given by sodium and potassium channels opening. It means that there isn't any information communicated between neurons in the shape of the potential action" (Rao, 2015).

A considerable number of different concepts of neuroscience are based on this idea. Moreover, they are the concepts opposite in meaning and spirit: discrete models counting the average speed of spike trains and wave models counting the probabilities of waves and their patterns. All of them treat the spike itself as a discrete event with the same shape. Some deprive it of any meaning and make it just one of the random events that, coming together, create order in a magical way. Others try to attribute sense to the set of these discrete elements.

Some ignore the meaning of the notes that create the music of the Mind, while others try to make identical sounds from the notes and calculate their speed or sequence. The latter are very close to considering the neural code as music since the sequences of code elements undoubtedly matter more than the tempo. But as long as they look at the notes themselves as identical, stereotypical pulses, the music will remain unheard, and musical notation is not written. The neural code will remain an enigma.

Some look at the oscillations of populations, ignoring the oscillatory nature of the very elements of these populations, which simplifies a lot as all the complexities of one neuron remain for the neuron itself. Others also leave these difficulties to the care of neurons but ignore the oscillations of the neuron itself and are skeptical about the oscillations of neural ensembles. They feel more comfortable analyzing the firing of individual neurons than with waves calling them epiphenomena, by-products of this firing. A single note has no meaning in neither concept. Not to mention the fact that pauses are not even considered.

Let's take as an example Predictive Coding Theory. Over the years, the authors of the model developed and promoted their concept, which, in principle, repeated an important idea of the nineteenth century about the predictive basis of the process (the famous hypothesis of Helmholtz about unconscious inferences). The authors describe it with complex math, but the idea is simple if you look at its essence.

The brain solves the inverse problem (identifying the causes of the observed from the observed). It is not a trivial task since reasons can be multiple. We are talking not only about the verbal level of consciousness with logical causal chain construction but the creation of any representation of incoming signals. Predictive coding means that the brain creates representations of signals and a model of reality as a whole as a set of possible causes (encoded parameters) of signals. New signals are processed based on created representations as known causes. Selection of the most probable causes is achieved by continuously minimizing error (discrepancies between sensory data and predictions).

It is a plausible and working hypothesis shared by many. Rao and Ballard are not the authors of the idea, but they are trying to give their model of the process and an algorithm for such a predictive function. The PAAL algorithm model proposed here is trying to solve the same problem. Therefore, the question is not in the general idea, which exists for a very long time (in some sense, Immanuel Kant and Thomas Aquinas spoke about the same), but in the details. The authors of the theory compiled a series of equations and diagrams, where "error neurons" and "prediction neurons" participate in the process. The former receive the signal and prediction parameters, add them and produce a residual error (difference). The latter receive the result from the former and correct the coefficients. All of them are connected in a chain with changing weights of the synaptic connection to regulate the transmitted code in the process of learning (adaptation).

The scheme is working, but at a very metaphorical level. Any variant of the code can be used in such a scheme. But what did the authors use as a basis for describing the physics and physiology of the process? The same average firing rate idea. There were many articles and discussions, formulas "flew" from one side to another, and a lot of fog was created. But an attentive reader would discover that the algorithm's authors demanded the impossible from neurons: create both a positive average firing rate and a negative one. How come? It is math; everything can happen in it. Physical and biological plausibility is not very important.

Of course, it was possible to rewrite the formulas of the algorithm and use only positive speed. It was done after the discovery of absurdity (Ballard, Jehee, 2012). But the problem is that absurdity was not a mathematical error. The logic of the proposed coding led to it. Such a linear code with one parameter cannot provide flexible and comprehensive coding of a potentially infinite set of both the signals themselves and their parameters, which are in constant dynamics. As a result, the very concept of speed became "flexible" in the model: it suddenly turned out to be negative.

No one could think of an excuse for such a violation of physical causality and plausibility. New algorithms were proposed, where the speed was initially limited by physically plausible positivity. And what was in these new versions, besides the "speed limit"? The same "good" old way of introducing auxiliary variables: coefficients appeared in the formulas that brought model neurons back to the reality of life and prevented division by zero errors, which stabilized the algorithm (Spratling, 2015). But there was one "small" problem: the neurons still didn't behave the way the model predicted.

If the original idea is wrong, then it will inevitably face a dead-end. But for some reason, the way out of the impasse is seen by many not in changing the idea but in using auxiliary variables, error "limiters," and other tricks of patching up "holes" in the explanatory base. All the same, down to the details, happened in standard theories of physics several decades before the similar events in neuroscience. And again, strange negative values for a parameter that cannot be negative by definition, and division by zero, leading to disappearance into nowhere.

It was more difficult for neuroscientists: they could not declare neurons as virtual particles, give them the go-ahead for time travel, disappearance and appearance by the magic of the operators of birth and annihilation. But they had access to the mathematical tricks of introducing "negative decision constraints" or "zero-order subtractions." They followed the path of theoretical physicists with a dippy hocus-pocus business of renormalizations.

Why was there a need for renormalizations in elementary particle physics? The absence of a truly physically plausible version of the mechanism for creating structures and organizing processes in them. The same happened in neuroscience.

If we talk about neurons as some kind of "shooting particles" and estimate the speed of their firing, then we cannot go into a negative solution to the equation: the speed is either positive or zero in a given coordinate system. Speed cannot be below zero, since then, the introduction of negative space and time is required. This trick is for special lovers of parallel universes and otherworldly places with curvatures of anything. But neuroscience still remains within the framework of the descriptions of the phenomena of this world and cannot afford to go into the flight of mathematical fantasy with negative speed solutions.

But the proposed model inevitably led to such an absurdity. What is the original problem with this version of the code? It is linear. The neurons themselves turn out to be linear operators, the network becomes linear, even if it includes forward and backward streams, and the description algorithm becomes linear. Linear models occupy a leading place, although they are such rough approximations to the described phenomena that they simply begin to contradict them. But if we consider a neuron as a kind of "shooting particle," then we inevitably come to the linearity of the model.

Real neurons are relaxation-type oscillators with physical time constraints. But in such models, the neuron must have time to give a sufficiently large series of shots so it would be possible to encode at least some information in the average speed. As we have already mentioned more than once, the system in real life simply does not have time for this. The spikes do not have either a negative speed or infinite speed. The only way out is to encode information in the spike itself.

If a neuron encodes with an average speed, then it should go in one direction or another from a certain reference point. No problem, says the linear model, this reference point is spontaneous activity without stimulus, and the average speed can be positive or negative relative to this base. That's all the "solution": simple arithmetic, where the positive value is added, and the negative value is subtracted.

But this is where problems arise. First, a neuron has to change its speed all the time, which means that there must be a constant change of neurotransmitters in synaptic contacts from activating to inhibiting, and this should happen almost instantly. With all the phenomenal capabilities of our body, this is biologically unrealistic. Second, if the spontaneous activity of the neuron is slow, then the model requires a negative average speed. It is pure mathematics, which, due to an initial error in the hypothesis, leads to a solution that contradicts physical reality if the observable parameters of the phenomenon are inserted into it.

If we come up with neurons with any speed and other magical properties, the average rate will always be positive, as it should be as a physical parameter. The problem is that the neurons become not physical but virtual. It is a vicious circle. Physicists gave an example of how to get out of it with a simple trick: come up with a virtual particle that can perform any miracles. There are many models that in great detail and mathematically correctly illuminate the activity of not real but virtual neurons. They are called "model neurons."

But we are talking about modeling actually observed cells and the dynamics of their potential, and not about virtual elementary particles, which no one has ever seen and, perhaps, will not see because they, most likely, do not exist. Theoretical physicists can invent backward time travels of particles and negative velocities. For neuroscientists, even pure theorists, it is more difficult to pass such fantasies as scientific concepts.

There is another version of the tempo code. Why don't neurons encode parameters with average speed relative to each other? For example, one neuron encodes a zero mark with its own speed, another encodes a positive change in a signal parameter with a higher rate, and a slow one encodes a negative change. This is the so-called biased encoding. Or like this: one neuron encodes a positive change in the parameter by a proportional change in the firing rate, and the other — a negative one, and both work in shifts when the parameter changes in one direction or another (two-cell encoding). If it is necessary to encode many parameters, neurons can be combined into many pairs encoding a change in one of the signal components. Why not?

Indeed, everything is logical. But again, we run into the same physical limitations of the speed of the neurons themselves with almost instantaneous speeds of signal processing and coding that they provide. This is the main stumbling block for any linear tempo coding scheme. It is a paradox: slow neurons work quickly. Are they ahead of themselves? This paradox is insurmountable if we look at neurons as particles with discrete stereotypical pulses. A similar dead-end is observed in the physics of elementary particles, leading to an endless search for a "particle of God" with a fundamental value of all parameters of which everything consists. But just as matter at the subatomic level consists of continuous dynamic notes of energy, cellular level matter is not "bricks" with the same and constant form.

Even if we think up a virtual neuron that can shoot fast enough to encode something with an average speed, or a chain of neurons encoding a signal with relative average speeds, such a code itself is so limited in information density that it cannot be a serious contender for the role of a neural code. Information about many constantly and rapidly changing parameters cannot be encoded in the tempo of identical notes. It should be contained in the note itself, in its characteristics as a wave packet with specific parameters, in combinations of notes, in sequences and durations, in pauses.

Neurons just have to accurately play their spike-notes, regulate their internal parameters and combine them with parameters of oscillations of other neurons. Moreover, they do not need to "worry" about providing constant firing. Their pauses also make sense because the phase structure can itself contain a considerable layer of information. It remains to make sure that both sounds and pauses are in place. This is ensemble play, synchronization.

The main mistake that led to the hypothesis of the neural code as modulation of the average firing rate was the logical error of mixing correlation and causality. When in 1926, researchers found a relationship between the load on a muscle and the rate of activation of motor neurons, they concluded that modulation of the firing rate is the basis of neural communication (Adrian, Zotterman, 1926).

And until now, in experiments, changes in the rate of activity of neurons in different areas of the brain under various conditions of motor and cognitive activity are taken as "proof" that neurons encode information in this way. The approach to neurons as the creators of discrete shots with the same characteristics remains the main road of neuroscience. If the spikes are the same, then the only variable parameter is speed. The question remains about the original premise: are the spikes the same?

The rate does change, but it can be very different under various conditions, even in the same experiment with the same stimuli. And then researchers have to average even the average speed (numerous experimental attempts fall into one heap) and look for meaning in such a mess, which was not there initially, except, of course, for the only obvious thing: a change in the speed of work of a neuron. Even the term "firing rate" itself is used in a different sense depending on the calculation procedure: as the average over an arbitrary observation time, as the average over several repetitions of the experiment, or both taken together.

But experiments show that neurons can produce the same average number of spikes when different stimuli are processed. It seems like this is enough to refute one of the most enduring myths in neuroscience about the tempo code, but the paradigms do not give up so easily. What happens to temporal patterns? They differ when coding different stimuli and repeat when coding the same stimulus in different experimental attempts (Rieke et al., 1999). Again, it seems that this should lead to a change in the model and shift the attention of researchers to the analysis of these patterns, but the usual method of saving the old model is to ignore the facts that contradict it. The model ceases to test reality and a hallucinatory-delusional state of the model's projection in isolation from the signals of the environment arises. It is the pathology of scientific knowledge.

In his book "Rhythms of the brain," neuroscientist György Buzsáki wrote: "Although stimulus-induced synchronization is often associated with increased firing rates of the responding neurons, ensemble synchrony can occur also in the absence of firing rate changes in individual neurons ... The most important message of these empirical observations in various cortical structures and species is that the information about the input can be embedded in the dynamic change of temporal synchrony even without an alteration of discharge frequency. And if no extra spikes are required to transfer information, no extra energy is needed ... Since the early days of sensory neurophysiology, it has been known that neuronal responses vary considerably from trial to trial. This variability has been traditionally thought of as "noise" that should be averaged out to reveal the brain's true attitude toward the input. However, the source of noise has never been identified and has been assumed to result from the brain's imperfections ... Extracting the variant, that is, brain-generated features, including the temporal relations among neuronal assemblies and assembly members, from the invariant features evoked by the physical world might provide clues about the brain's perspective on its environment. Yet, this is the information we routinely throw away with stimulus-locked averaging" (Buzsáki, 2006).

The actual variability of the patterns is thrown away as noise, and the tempo is averaged over the different measurement attempts. This approach does not initially consider the frequency, phase and amplitude characteristics of neuron oscillations and the dynamics of ensemble patterns, and, as a result of averaging, what could be extracted from measurements is completely lost. Notes, melodies, harmonies and rhythms of the brain are ignored. The nuances of the music of the Mind drown in such "analysis." The researchers do not set themselves to find these details since they simply do not exist for such a code model. But the model does not correspond to the realities of the system.

Again, we have to emphasize: the system does not have time to create messages from the average number of spikes when encoding signals and counting these spikes when decoding. The system is forced to develop fast and information-rich code to survive. It cannot do without nuances. When scientists in the laboratory calculate averages, they proceed from logic: if the basis of the code is discrete, identical pulses, then you need a sufficient statistical sample to create an average speed of such pulses. The logic is flawless, but the problem is with the initial premise: the hypothesis that the neural code is the average number of peaks in the phase of the potential oscillation, and the rest of the dynamics does not carry any information.

The rate code hypothesis is sometimes referred to as the "frequency code." But this is a severe mistake of confusing and substituting concepts. If we take the continuous physical cycle as a discrete "shot," then the rate of firing per unit time is not the frequency of the oscillation but the average rate of discrete events. We removed the entire development of the wave, squeezed it to a peak, and counted the number of peaks per unit of the nominal time. Then we also averaged this amount, which changes during the observation, over an arbitrarily chosen time interval and over different experiments to obtain a certain general meaning.

But if we remember that oscillation is a continual process, then it has a stable phase space in which the dynamics of phases unfold. Even one cycle has internal dynamics, and the frequency of oscillations will be the alternation of cycles with a sequential change of parameters.

Let's put ourselves in the place of the receiving neuron (decoder). Do not forget that it is an oscillator that must tune in to incoming energy fluctuations. If the meaning of incoming signals is in the number of phase peaks (spikes) in some period, then it will have just to wait and accumulate a statistical sample, as scientists do in a laboratory. But they have time and money (energy) for this if there are enough funds and grants. The neuron has neither extra time, much less free energy.

Now imagine that there is information even in one cycle of the incoming oscillation: the initial phase and its place in relation to the phase of the decoder itself, further dynamics, "distance" (time) between phases, pattern and form of even one cycle. Then the neuron doesn't have to wait for anything, accumulate and average anything. But it has to understand this message with all its nuances, to be tuned in to perceive these parameters. And it will "count" the frequency of the incoming oscillation relative to the clock of the system or population, of which it is a part, and not in seconds and milliseconds (external measure). If there is one cycle in this measure, then this is one frequency, two — another, and so on. The entire frequency range must be inscribed in the base pulse at the reference clock frequency so that elements of the system in different places and various states can relate messages to each other and read them. And if the clock cycle of the system meets its adaptive needs, then there will be enough information in it even in one oscillation per measure. But in the average number of phase peaks in some arbitrary unit of time used by an external observer, there will be absolutely no information.

And for the observers themselves, there is no information in this, except for the calculated average number of peaks. Just imagine that we are trying to understand music, and for this, you measure the tempo as the number of notes in an arbitrary unit of time but ignore the bar structure of this music. We average this amount within one piece, then mix it with another performance of the same piece at a different tempo, then again and again. We average all the data obtained and try to draw a conclusion about the meaning of the music.

And all this time, the meaning was in melodies, harmonies and rhythms. They were the same but played at a different pace while maintaining measure and structure. We did not want to hear them initially (conceptual error) and completely drowned them in subsequent attempts to understand the meaning (measurement and analysis error arising from a conceptual error). But on the other hand, it was easier for us: why bother with notes, melodies, harmonies and rhythms, when there is such a noticeable and measurable parameter as tempo? We feel uncomfortable with the details of the oscillatory and wave processes, but we are ready to patch up conceptual holes with arbitrary variables and other tricks.

Some say that although there are complications at the level of the hierarchy of the neuron, they should not even be considered because the population and its activity should be the correct element for the model. Others say that spikes have no complexity, and they just have to be counted as identical shots in a row for a specific time. For all the apparent differences in approaches, they have the same essence: the spike itself is not important for them. Perhaps they are correct, and the spike of one neuron is not important for the brain itself? And there is no information in it, except for the very fact of the presence of a spike, as a discrete shot according to the "all-or-nothing" principle with a clear threshold and standard values of the parameters of this shot?

A standard recording of neuron activity looks like the same sticks, spread out with different densities along the time axis:

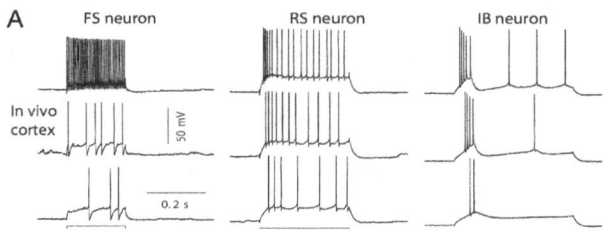

Shown here are in vivo recordings of different variants of firing patterns called fast spiking (FS), regular spiking (RS), and intrinsically bursting (IB) (Rulkov, Bazhenov, 2008).

They look like beautiful and identical spikes made to the order of supporters of the firing rate coding model: it remains to count the amount per unit of external time, associate it with the stimulus and understand how this stimulus is encoded in this amount. It has been done for decades by the proponents of this approach to neural code.

And again, the question arises: maybe this is how it should be? Let's try now to forget all our previous arguments about the shortcomings of the very concept of the tempo code and just focus our attention on this issue. Maybe the spikes are really such even "sticks," and there is no complication at this hierarchy level? Should we just simplify and forget or admit that there is complexity but leave it to the care of the neurons themselves? Let them play their various notes and parts, but we only need the overall music of the orchestra and the general parts of "strings," "brass," and other populations. And, indeed, why should an inexperienced listener know all the subtleties of musical notation, all the nuances of each note, if his task is simply to enjoy the music?

This approach has a right to exist if we are just users. We have lived for thousands of years without any idea that any neurons are working there at all. Now we know that they are there and are doing something. Let them do their own thing, and we will enjoy this music of the Mind without analyzing the subtleties of the musical notation.

Everything would be so if this music sounded perfect and harmonious all the time. But the problem is that there are problems that we call disorders of the Mind. And if we want to understand and correct them, we will not be able to do this without mastering the laws of this code and reading the musical notation of the brain.

In the same way, we would not be able to correct disorders of the music of the heart if we did not learn its patterns. After all, if our cardiovascular system worked without failures, there would be no need to understand its music. We would not even have to listen to it playing in the background. But no, we have been listening to this music for a very long time, and it is no coincidence that the stethoscope is considered a symbol of belonging to the medical profession. And there are outstanding results and progress in understanding this music. The stethoscope has become mainly a symbol, as the variety of diagnostic and therapeutic equipment has increased dramatically. It is all aimed at the sounds of music of the heart and other organs.

Neuroscience is not a single entity that sits in a laboratory and looks at neurons. This is a vast number of people with a considerable number of opinions. But there are tendencies in which these opinions merge. Now we are considering one, albeit a powerful trend. It can be conditionally designated as follows: we do not need musical notation. We can take a variety of notes, reduce them all to a shot and write them down in the form of the number of these shots in an arbitrarily chosen unit of time, which is convenient for us. And safely forget about the possible complications: the fact that these notes have their individual parameters, and the measure of the system may be different. Why do we have to know the bar if we do not study the rhythm but calculate the tempo? The temporal structure as the duration of notes and their location on the clock frequency grid is not required to determine the tempo. It is enough to fix the peaks of the bursts and count their number at any arbitrary time interval.

The problem is that we can do this indefinitely, improving our technologies for calculating the rate, but it will be running in place. And the enormous amount of information received will not make any sense. It will not reveal the pattern of the code, its semantic content inherent in the musical score of the music of the Mind. And we will make the usual mistake: wishful thinking, projecting a model that is divorced from the meaning of the introjected signals. Our notation of this music will look helpless. Such a record of similar spikes is not even neumes. All notes are reduced to one tick repeating at different speeds without any identifying signs.

But is it possible that the brain actually produces such identical notes? The fact of the matter is that it does not. And neuroscientists, even the fiercest proponents of firing rate, know this. But, as already noted, this is more comfortable: we will forget about the complexity of oscillations of the neuron and look either at population oscillations, or we will not think about oscillations at all but count discrete ticks.

Let's take a close look at what is measured and shown in the firing pattern charts we saw above. To measure signals in vivo, researchers use different variants of invasive technologies with electrodes immersed in brain tissue. For obvious reasons, this research option is limited to using either animals or humans who undergo surgical intervention and gave consent to the experiment. With animals, it is difficult to draw conclusions about the links between the data being taken and internal cognitive activity. In the absence of direct self-report of such subjects, it is necessary to draw indirect conclusions from the behavior. Human studies are limited by the time of the surgery, the place of intervention (only those zones that are in a pathological state and are the target of the operation can be examined), and the drug effect on the brain.

But these technologies have advantages: they measure signals with temporal resolution at the level of milliseconds, which corresponds to the time of neurons' operation; good spatial resolution (within 1-2 mm); they measure the signal directly from one neuron or the potentials of a narrowly localized field; have a better signal-to-noise ratio since there is no such dependence on interference introduced by other organs and movements of the subject; allow measuring a wider range of frequencies. In addition to passive measurement, they can be used for electrical stimulation of tissues, making it possible to find functional connections between populations and the relationship between a given zone and behavior, internal perceptual-apperceptive processes.

Such technologies allow making measurements at a much finer level than EEG and MEG. But the question arises: what is measured? It is understood that the electrical activity of the brain. But which one? The tip of the electrode measures what is around it. Many events are happening at the intracellular and intercellular levels. If we measure the activity of one neuron, then this is a conditional formulation denoting a certain "soup" with a large number of ingredients. All activity is reduced to transmembrane fluxes measured by an electrode in the extracellular environment. It includes synaptic fluxes and ion fluxes in all channels along the membrane, from presynaptic dendrites to postsynaptic axons.

The electrode can only measure the potential difference between the points, and the central question becomes what kind of points we measure and what process between these points we register. The term "local field potential" (LFP) is used to denote the readings of the electrodes. It shows the approximate nature of the measurement and reflects the usual use of the magic word "field" when there is a loss of the physical meaning of what is happening.

The authors of one review even called this term malapropism — the lexical error of replacing one word with a similar sound to another. They wrote: "The term 'local field potential' (meaning an electric potential (V_e)) is a regrettable malapropism, but we continue to use the term LFP because it is familiar to most neuroscientists" (Buzsáki, Anastassiou, Koch, 2012).

But this is not a stylistic error. As usual, the term, even incorrect, reflects the model and approach to the phenomenon. If we call something "local field," then this gives us the illusion of understanding and even the illusion of explaining the physics of the process at the local level. And when we draw representations of such a phenomenon in the form of graphs of the field potential and analyze them, we can average and smooth the results even imperceptibly for ourselves. And a "miracle" happens: the details of the oscillatory process initially lost during the measurement due to objective capabilities of the equipment completely disappear during further analysis due to the subjective factor of the model. The complex oscillatory process with its frequency, phase and amplitude parameters suddenly becomes identical "sticks" on the graph. The notes of the neuron's music just disappear. And then, of course, we don't need musical notation to describe them.

But in real physics and physiology of the process at the level of one neuron, the "field" is a synchronization pattern and a superposition of all processes in ion channels throughout the cell. Why did a neuron acquire so many complex membrane channels at all, each one as an oscillator with its parameters, if the task is only to fire the same "shot" at different speeds? If it is a linear accumulator with a clear threshold, there is no need for such complex logistics and logic. But the fact is that the whole complicated mechanism of neural ion channels exists in order to fine-tune the parameters of a neuron as a relaxation oscillator to synchronize it with the rest of the ensemble.

Each neuron and the entire brain as a whole are a hybrid analog-digital chain. There are no even spikes, rows of digital 1s, as drawn in many models dealing with neural code. The notes of the neural code are not the same, they have their own continual oscillation parameters, but at the same time, they are also discrete elements of the general continuous process.

The review authors wrote: "The characteristics of the LFP waveform, such as the amplitude and frequency, depend on the proportional contribution of the multiple sources and various properties of the brain tissue ... In addition to the magnitude and sign of the individual current sources, and their spatial density, the temporal coordination of the respective current sources (that is, their synchrony) shapes the extracellular field" (Ibid).

Some ion fluxes (for example, Na^+) can be intense but short (less than 2 msec) and simply do not fall into the frequency band of the electrode measurement. Powerful and long-lasting dendritic flows of Ca^+ occupy a significant share in the final picture obtained by the electrode. A considerable number of processes in different channels are not direct synaptic events leading to an action potential and its transmission, but these currents affect the inherent vibrational characteristics of the membrane.

"Voltage-dependent resonance and oscillations at theta frequency have been described in principal neurons of several cortical regions. By contrast, perisomatic inhibitory interneurons have a preferred resonance in the gamma frequency (30–90 Hz) range. Because resonance is both voltage- and frequency-dependent, its impact on the magnitude of the extracellular field can vary in a complex manner" (Ibid).

The vibrational characteristics of different fluxes can differ within the same cell and between different types of neurons. For the electrode to catch it in its "local field" net, the fluctuations of the potentials of different flows in one neuron and in those adjacent to it must coincide in time. Or, more precisely: everything that falls into a given place and at a given time of measurement merges into the readings of a particular electrode.

These can be input signals on dendritic trees, signals in the soma of a neuron, axonal flows, carrier basic pulse of the membrane potential, synaptic action potential, afterhyperpolarizations (AHPs), direct flows of ions in electrical synapses (gap junctions), and even changes in the membrane potentials of neighboring glial cells, which are not directly involved in the neural code.

In addition to all this complexity of actual flows, which is reflected in the LFP of the electrode only as an average value, topography has a significant impact on the readings. For example, in the cortex, the dendrites of pyramidal neurons are parallel to each other, and the incoming afferent pathways lie perpendicular to the dendritic axis, which is ideal for superimposing signals from active dipoles, and therefore the LFP signal from the cortex is powerful. But the cortex gyri have different concentrations of dendrites on the concave and convex sides, which affects the power of the signal received by the electrode.

In the hippocampus, the bodies of neurons are displaced vertically relative to each other, which leads to mutual suppression of flows from the neuron soma and the dendrites of neighboring neurons. As a result, in rodents with a small size of the hippocampus and only a few rows of pyramidal neurons, the LFP signal is quite pronounced, but in primates and humans, it is no longer so. The result is a paradoxical situation: although the size of the hippocampus itself and the complexity of its structure increase, the measured signal becomes less and less pronounced.

"Extracellular action potential (EAP) recordings form one of the primary means for studying the activity of the intact brain ... Typically, EAP recordings are used only to determine whether and when neurons have spiked, under the assumption that the actual waveform of individual EAPs does not convey any information. At the same time, average EAP waveforms are known to exhibit a range of characteristic features when observed on a millisecond timescale, and these variations can be used to distinguish between different neuronal classes as well as individual neurons within classes. However, there have been few attempts to systematically study the causes of the variability in EAP waveforms either through experimental work or through computer modeling. The paucity of studies probably results from both insufficient data and inadequate techniques with which to model EAPs in a meaningful way" (Gold, Henze, Koch, Buzsáki, 2006).

Indeed, there are objective factors for such a state of affairs when the actual physics and physiology of the wave and oscillatory process is ignored, and discrete even spikes are drawn. Of course, the inadequacy of technology also gives rise to insufficient data. Still, the basis is the same uncomfortable state of many neuroscientists when faced with the reality of the dynamics of oscillatory processes from the intracellular to the population level. It is customary to consider discrete spikes, although they are not quite spikes (sharp sticks) and not quite discrete.

The authors of the article cited above made a computer model of EAP generation, changed different parameters, compared the behavior of simulated neurons and real in vivo data. Their study showed that the location of the electrode significantly changed the resulting waveform. But most importantly, the shape varied with fluctuations of such parameters as the composition of different ion flows and distribution of ion channels.

The authors concluded: "Our findings suggest that accurate and high-resolution monitoring of EAP can provide information about alteration of conductance densities in single neurons as a function of state changes and plasticity ... Our results imply that by monitoring the waveforms, EAPs can provide access to this valuable information" (Ibid). Indeed, until we begin to approach each spike as a note with a different waveform, with different internal parameters, we will not receive valuable information about the music of the Mind and its musical notation (code).

This is how the action potentials are drawn in standard models:

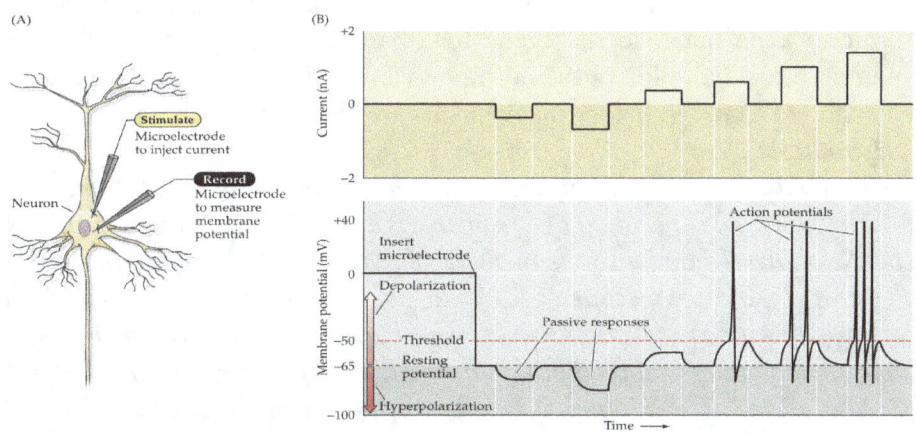

Purves et al., 2012

A very subtle technique is used to register individual spikes in vitro. Still, the basic idea is simple: two electrodes are connected to a neuron in a section of brain tissue, a current is applied to one, and the other measures the dynamics of the membrane potential. When a constant signal with increasing power is applied to the input electrode, no linear dependence of the amplitude of the action potential on this signal is observed. There is simply an increase in the number of spikes.

The result looks like sharp spikes or identical discrete shots with their number increasing with the applied current. This leads the researchers to the conclusion that the neuron works like a flush tank: it accumulates and drains. The incoming signal is thus encoded by the rate of accumulation and drainage, the firing rate. The action potential itself looks like an all-or-nothing event: it either happens or not, 1 or 0. Zeros (pauses) are not taken into account in the analysis, and the code itself is represented as a set of ones with a certain speed.

At a first and superficial glance, it seems that this is the whole truth of neural communication. Indeed, there is a burst as the peak phase of the membrane potential oscillation. The fact that the bursts become more frequent as the current increases is also a fact. But this is part of the truth and a tiny one. The devil, as always, is in the details. These details depend on the degree of resolution. To understand whether this picture reflects reality or not, we have to go to the temporal level of the neuron itself.

If we increase the resolution along the time axis, then the picture changes dramatically. A spike from a sharp tick turns into a wave with a shape that varies depending on the state of all parameters and the logistics of the entire complex chain of one neuron as a signal transducer. The "one" of the action potential turns out not so simple and discrete but a complex analog process. And then it becomes clear that the neuron is not only and not so much a linear "flush tank," but a filter-oscillator, the impulse response of which is a dynamic variable. And the threshold is not at all fixed. In this sense, there is simply no threshold. However, there are ranges of the incoming signals parameter values at which a neuron exhibits certain changes in the parameters of its oscillation. The impulse response of a neuron as a filter has specific characteristics and settings. The "dam" of a neuron is not a fixed concrete structure but a flexible and delicate mechanism for adjusting a neuron as a musician in the orchestra of the brain. Neuron oscillations have dynamics and a complex phase portrait of a non-linear nature. By the way, even in a concrete dam, regulatory mechanisms are also necessarily in-built.

Hypothesis:

The continuity of the process is a natural physical and physiological phenomenon of the operation of a neuron. Its ability to synchronize in an ensemble depends on the dynamics of the membrane potential. A change in the spike shape can have different consequences in various frequency ranges and population activity patterns as traveling waves. The dynamics of voltage changes become either an activating or an inhibiting factor (a push-pull mechanism) depending on the phase of the oscillation cycle and the frequency of the interacting oscillators. Phase and frequency coupling of neurons depend on the dynamics of voltage during the action potential and in-between.

It is mathematically quite feasible to introduce variables that could describe the change in voltage depending on the rate of formation of the peak potential, the width of the peak phase (spike) itself, and the relaxation period. All these parameters and their combinations affect the phase and frequency coupling of neurons in the ensemble.

The total number of equations and variables even for describing a network of two interacting neurons will be huge. But it is also possible to reduce the model to a small number of parameters and analyze synchronization stability depending on their dynamics. There are attempts at such an analysis (see the review in the article "Mechanisms of phase-locking and frequency control in pairs of coupled neural oscillators," Ermentrout, Kopell, 2000).

Let's pay attention to what kind of signal the researchers give as input when analyzing the neuron's action potential dynamics. Does this signal make any sense to the neuron? The answer, to the dismay of many generations of researchers, is negative. Yes, an electrical signal is sent to a neuron. Yes, electricity is the neural network's communication medium. But the question was about the meaning, and meaning is code, pattern. For a neuron to understand and respond meaningfully, it is necessary to speak with it in the same language and give it sensible messages. To receive a rhythm, you have to send a rhythm. What do researchers supply to a neuron? The flow.

The word rhythm comes from the Greek "rhuthmos," which is derived from the verb rhein (to flow) and the suffix mos, which gives this word the meaning of not just an unstructured flow, but a definite form in time and space. But the researchers just pour a stream onto the neuron and look: will it choke or not? Will it drain or not? It can drown if completely flooded. But within some limits of the flow rate, it will simply desperately "spit-out," it will integrate and fire on the all-or-nothing principle.

The researchers will look at the result of this "spitting" and think that this is the neural code. But since they will not find a particular pattern in these "spits," they will say that the neuron is kind of "stupid" and "straightforward": its spikes are random, they have a higher or lower speed, and it contains all the meanings that we call consciousness. How simple: the code was found without any particular difficulties, and if there are any, then we will leave them at the level of the neuron itself, let it deal with them on its own. It will deal with them, but have we dealt with the neural code in this way? No.

In his book "Dynamical Systems in Neuroscience: The Geometry of Excitability and Bursting" entirely devoted to the details of dynamics at the neuron level, Eugene Izhikevich wrote: "Most introductory neuroscience books describe neurons as integrators with a threshold: neurons sum up incoming PSPs and "compare" the integrated PSP with a certain voltage value, called firing threshold. If it is below the threshold, the neuron remains quiescent; when it is above the threshold, the neuron fires an all-or-none spike and resets its membrane potential. To add theoretical plausibility to this argument, the books refer to the Hodgkin-Huxley model of spike-generation in squid giant axons … The irony is that the Hodgkin-Huxley model does not have a well-defined threshold, it does not fire all-or-none spikes, and it is not an integrator but a resonator, i.e., it prefers inputs having certain frequencies that resonate with the frequency of subthreshold oscillations of the neuron" (Izhikevich, 2007).

Even such an old and simplified model spoke about oscillations and waves and not just about discrete spikes. But over the past decades, this point has somehow disappeared from the field of view of standard theories of neuroscience. Let's take the same "Neuroscience" textbook that we have quoted many times. You will not find a single word about oscillations, waves, and synchronization in the entire introductory chapter entitled "Electrical Signals of Nerve Cells."

The discomfort in contact with non-linear oscillatory dynamics leads to ignoring facts and even outright manipulation with data. Izhikevich wrote: "The drawback is that the measurement procedures may not be accurate, that the parameters are usually measured in different neurons, averaged, and then fine-tuned (a fancy word meaning "to make arbitrary choices"). As a result, the model does not have the same behavior as one sees in experiments ... Many scientists, including the author of this book, refer to these neural models as being 'spiking models.' The models have a threshold, but they lack any spike-generation mechanism, i.e., they cannot produce a brief regenerative depolarization of membrane potential followed by a slow hyperpolarization. Therefore, they are not "spiking models"; the spikes in the next two figures, as well as in hundreds of scientific papers devoted to these models, are drawn by hand ... Figure 8.1: Leaky integrate-and-fire neuron with noisy input. The spike is added manually for aesthetic purposes and to fool the reader into believing that this is a spiking neuron.

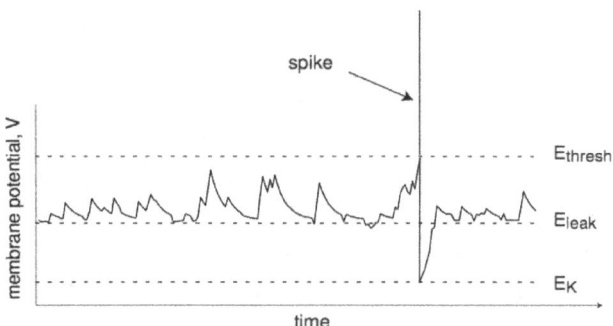

Izhikevich, 2007, www.izhikevich.com

The leaky integrate-and-fire model is an idealization of a neuron having Ohmic leakage current and a number of voltage-gated currents that are completely deactivated at rest. Subthreshold behavior of such a neuron can be described by the linear differential equation ... Since the shape of spike is not simulated, all spikes are implicitly assumed to be identical in size and duration. A stereotypical spike is fired as soon as V=Ethresh, leaving no room for any ambiguity ... The neuron can continuously encode the strength of input into the frequency of spiking ... In summary, the neuron seems to be a good model for an integrator. A closer look reveals that the integrate-and-fire neuron has flaws ... Finally, the integrate-and-fire model is not a spiking model: technically, it did not fire a spike in Fig. 8.1, it was only "said to fire a spike," which was added manually afterwards to fool the reader. Despite all these drawbacks, the integrate-and-fire model is an acceptable sacrifice for a mathematician who wants to prove theorems and derive analytical expressions. However, using the model might be a waste of time" (Ibid).

Indeed, over the decades, within the framework of this approach, thousands of scientific publications have been written, and probably hundreds of theories and models of neural code based on idealization and drawing of clear and identical spikes have been created. And all of them are not so much to "fool the reader" (although such motivation is not excluded), but more to fool themselves. It is wishful thinking: the neural code consists of the identical discrete spikes which we created ourselves, but they are so beautiful and even, and the most important thing we can count them and crack the neural code. Decades go by, and perhaps Izhikevich is right: there were many theorems and theories, many beautiful formulas and exercises for mathematicians, but maybe it was just a waste of time (and money).

I would like to soften the blow inflicted by Izhikevich. A lot has been acquired during this time, and a negative result is also a result. It remains to accept it as a mistake and understand in which new direction to move. Until such a paradigm shift takes place, we may carry a "suitcase without a handle," not wanting to throw it away because so much "good" old stuff has been accumulated in it. However, part of this burden may not just be excessive and interfere with further movement, but stop it or send it in a vicious circle instead of moving along the lemniscate with a central functional unit for assessment and correction. Indeed, in science, as in any other manifestation of the process of cognition, the one who does not notice or forgets mistakes condemns himself to their repetition.

Let's see what kind of picture of the dynamics of the potential of a neuron is obtained if we widen the time axis:

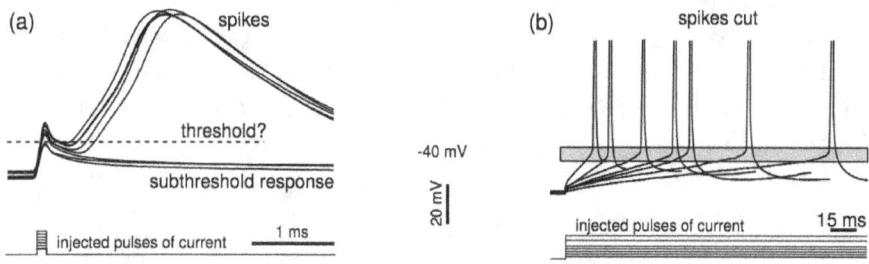

Izhikevich, 2007

It is a chart of the actual pyramidal neuron's potential in the cortex of a rat. The figure on the left clearly shows that the spike as a discrete event simply does not exist. There is no "sharp thorn" or "stick," and there is no definite threshold. The action potential was called a spike a long time ago, and this is another, albeit traditional, but misleading term. The action potential is a wave, a continuous process with different phases. Neurons do not shoot with sharp spikes; they vibrate softly with waves.

Is there a clear voltage value at which the neuron fires a spike? Izhikevich wrote with humor: "If you find one, let the author know! ... The firing threshold, if exists, must be somewhere in the shaded region, but where? Where does the slow depolarization end and the spike start? Is it meaningful to talk about firing thresholds at all?" (Ibid).

But if we narrow the time axis and even touch up the spikes to even sticks with the same height, then a picture appears that is familiar to generations of researchers. And we find something that is not really there. Izhikevich noted: "The notion of a firing threshold is simple and attractive, especially when we teach neuroscience to undergraduates. Everybody, including the author of this book, uses it to describe neuronal properties. Unfortunately, it is wrong. First, the problem is in the definition of an action potential ... Suppose we define an action potential to be any deviation from the resting potential, say by 20 mV. Is the concept of firing threshold well-defined in this case? Unfortunately, the answer is still NO" (Ibid).

But there must be a threshold as a bifurcation point of transition of oscillations from one phase to another? The fact of the matter is that if we begin to consider a neuron as a non-linear dynamic system and study phase portraits, then transitions and bifurcations must be there, but they are not some rigid and precise barriers. The problem with integrate-and-fire models is that they approach neuron activity as a discrete, piece-wise function. And then the bifurcation becomes, as Izhikevich writes, weird. By the way, almost all of his book is devoted to studying phase portraits of neurons, types of natural bifurcations, and the dynamics of neurons as oscillators. He believes that "a good neuronal model must reproduce not only electrophysiology but also bifurcation dynamics of neurons" (Ibid).

When we consider the process discretely and think in terms of linear accumulate-drain schemes, we imagine a certain threshold over which the flow must go. But this is not just an idealization of the actual mechanism, but a simplification leading to a dead-end. The model is increasingly detached from real phenomena and becomes a phantom. So, does it make sense to talk about the threshold? Yes, but in the meaning of the bifurcation point as a dynamic parameter of the qualitative change in the phase portrait of the system.

Neural communication is a continuous process of interaction of energy oscillations. If we are talking about a threshold, we should consider that the whole point of the process of fine regulation of all settings of the system called a "neuron" is to dynamically adjust its impulse response as a filter in the process of signal transduction and transmission of energy-information. And, of course, there are settings values at which the phase transition occurs. But firstly, they are dynamic, and secondly, the process itself is continuous and with many degrees of freedom. If it all boiled down to the strength of the flow that the neuron can "swallow and not choke," then the process of consciousness, as the creation of complex frequency and rhythmic patterns, would be impossible.

What is surprising is not that a mistake was made when choosing a standard model, but that it still lives on, despite its internal contradictions and inconsistency with the observed phenomena. Although if we remember the history of science in general and physics in particular, then this state of affairs will also cease to be surprising.

But the neuron stubbornly refuses to work the way the standard models prescribe it. Suppose we do not send a simple monotonous signal to the input or signals of different amplitudes, as is often done in experiments, but try to imitate a particular pattern. It turns out that when a signal has a specific high frequency, the neuron does not produce any spikes.

How come? We give a strong and frequent signal, the neuron must accumulate it and fire if it is a linear integrator, but it is silent. Is it saving up? No. It simply dumps the extra stream but does not give out any information to the outside. Why? Because a signal with this frequency and such a pattern does not make any sense to it. But if we send a signal with a lower frequency, the neuron will immediately "speak." If we lower the frequency more, it will be silent again.

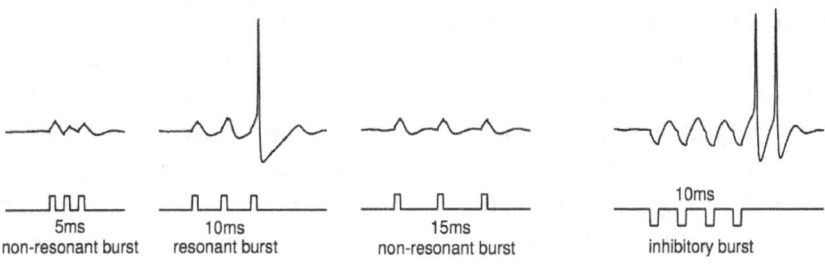

Izhikevich, 2007

Another type of neuron will react to different patterns of incoming signals and their frequency. "Thus, seemingly inessential differences in parameter values could result in drastically distinct behaviors" (Ibid). Why is this happening? Because besides the accumulation of flows, the neuron also perceives them as an oscillator, as a synchronized self-oscillating system. And if the incoming signal matches the settings of its impulse response, then there will be a result in the form of an outgoing activity pattern.

Usually, this property of neurons is called resonance, meaning that the frequency of the incoming signal must coincide with the natural frequency of the neuron. But the synchronization process is not reducible to resonant phenomena as coincidences of the frequency of the external influence and the natural frequency of the oscillator. There are many degrees of freedom and sync regions beyond simple unison of resonance.

Standard theories in neuroscience postulate two main points when talking about neural code: the basis of the code is similar spikes; the mechanism for generating code patterns is a cumulative ("integrating") function. But, as we can see, neither concept reflects the realities of the observed phenomena. The action potentials are not at all the same spikes but waves with different shapes. The accumulation does not necessarily result in a spike pattern, and in some cases, the input does not generate a single action potential.

The neuron and the processes in it are a much more subtle and non-linear mechanism, and if one can call it integration, then it is not at all in a sense used in the integrate-and-fire (accumulate-drain) concept. Integration, most likely, will be synonymous with synchronization because we are talking about the unification of many energy flows with different parameters into a single form of an outgoing action potential as a continuous wave structure carrying information in a set of its parameters, and not just in the very fact of its existence.

As long as we perceive the integration in a neuron as a kind of addition of flows up to a certain threshold, beyond which they are no longer different flows but a single outgoing one, everything seems clear to us. Izhikevich writes that this is a simple and convenient concept but it is erroneous. At a certain level of generalization, the concept of simple addition and a threshold, as a particular value of some parameter, is working one and reflects part of the truth. But again, the devil is in the details. When we plunge into the subtleties of the process, expand the resolution characteristics in space and time, the realization arises that the metaphor of the threshold in its usual meaning of a rigid barrier does not work.

But is it necessary to abandon such a metaphor? No. We just have to expand and deepen so that it does not conflict with the observed phenomena. Otherwise, a thoughtful student will inevitably ask the professor: how is it that we pour in and pour in, but it still doesn't pour out? Or so: why, when we pour in with one frequency, it pours out, and at another does not pour out, no matter how much we pour in? And so on. And if the professor imagines this process as a linear pouring in-out, as the work of a flush tank, he will have difficulty answering.

But suppose we remember that we are studying energy oscillations and their interaction. In that case, the questions disappear by themselves: the professor should advise the student to go through the oscillations and synch physics course first. This is still a hypothetical dialogue because there are no courses in which they would first study the physics of universal processes and then move on to physiology and details of specific processes in the nervous system and the body as a whole.

For example, without understanding the physics of sync, a researcher is bound to look at subthreshold activity as random noise. But for a neuron, most likely, it makes a lot of sense. Otherwise, why such a waste of energy? Ion pumps do not drive flows back and forth for free: they use up to 2/3 of the cell energy. But the living matter does not waste energy without a purpose. We can even say that it is the main distinction between living and non-living matter. So, what is the aim of all this hustle and bustle usually called "resting membrane potential"? Is it really rest? It is the basic synchronizing pulse: all frequencies, all incoming and outgoing notes are superimposed on it. This pulse is not random at all but meaningful. Moreover, being a pacemaker, it should be as stable and periodic as possible.

One study clearly showed that there is coherence between subthreshold oscillations of different neurons. The authors recorded intracellular oscillations in neurons of the visual cortex (V1) of a cat and found a significant correlation of spontaneous fluctuations in membrane potential in pairs of neurons. They note that "this synchronization was not dependent on the occurrence of action potentials, indicating that it was not caused by mutual interconnections. The cells were synchronized continuously and rather than for brief epochs" (Lampl, Reichova, Ferster, 1999).

Thanks to the basic subthreshold pulsation of cells synergy-synchronization arises, forming a pattern of action potentials as a neural code. It is the baseline and therefore does not depend on action potentials. Everything is exactly the opposite: patterns of activity are superimposed on it. And again, it is necessary to emphasize that the pattern (rhythm) includes not just the potentials themselves, their durations and sequences, but also the corresponding durations and sequences of pauses.

As Izhikevich accurately noted: "Is zebra a black animal with white stripes or a white animal with black stripes? This seemingly silly question is pertinent to every bursting pattern: Does bursting activity correspond to an infinite period of quiescence interrupted by groups of spikes or does it correspond to an infinite spike train interrupted by short periods of quiescence? Biologists are mostly concerned with the question of what makes the neuron fire the first spike in a burst and what keeps it in the spiking regime afterwards. The question of why the spiking stops is often forgotten" (Izhikevich, 2007).

Let's ask the same "silly" question about music: is it a set of sounds interrupted by silence, or silence interrupted by sounds? At first, the answer seems obvious: if music sounds, then it is a stream of sounds. What kind of silence are we talking about? This zebra is entirely white (or black). But it's not that simple. Even if we are talking about a part of one instrument, it always consists of sounds and pauses. Especially when it comes to a complex orchestra playing: the musicians play at the right time following their part where each note has a specific duration. The same goes for pauses.

So, the question again: one part — is it sounds interrupted by pauses, or pauses interrupted by sounds? The question is really meaningless because it suggests hierarchy where there is none. From whatever side you look at this "zebra," it contains both, and the meaning is precisely in the presence of both. The real zebra is not saved by white or black stripes but by the presence of a white and black pattern. Music is made off sounds and silence. It sounds paradoxical, but it is a physical fact.

The concept of synchronization as simultaneous firing says that if presynaptic neurons' spikes come together, they will create the necessary responses in the postsynaptic neuron as the target where this joint firing converges. If they do not coincide, then there will be no response or too weak. This is the basis of the neural code in many models.

"An implicit assumption here is that the axonal conduction delays are negligible or equal. A careful measurement of axonal conduction delays in the mammalian neocortex showed that they could be as small as 0.1 ms and as large as 44 ms, depending on the type and location of the neurons. A typical distribution of axonal propagation delays between different pairs of cortical neurons is broad, spanning two orders of magnitude. Nevertheless, the propagation delay between any individual pair of neurons is precise and reproducible with a sub-millisecond precision. Why would the brain maintain different delays with such a precision if spike-timing were not important? The majority of computational neuroscientists discard delays as a nuisance that only complicates modeling. From a mathematical point of view, a system with delays is not finite- but infinite-dimensional, which indeed poses some mathematical and simulation difficulties. In this paper we argue that an infinite dimensionality of spiking networks with axonal delays is not a nuisance, but an immense advantage that results in an unprecedented information capacity. In particular, there are stable firing patterns that are not possible without the delays" (Izhikevich, 2005).

To illustrate, Izhikevich draws a simple diagram:

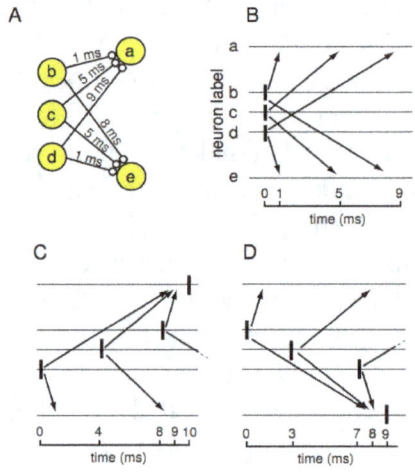

Izhikevich, 2005

The first figure A shows model neurons with different conduction delay times, and postsynaptic neuron (a) receives signals from several neurons. Figure B shows that when firing synchronously, the spikes will come at different times. It is simply ineffective and unrealistic if the task is to obtain an integrated result. If there is a temporal pattern, then the signals converge in time (Fig. C). And another pattern (Figure D) will cause the signals to converge in another neuron (e). "We see that depending on the order and the precise timing of firing, the same three neurons can evoke a spike either in neuron *a* or in neuron *e*, or possibly in some other neuron not shown in the figure. Notice how the conduction delays make this possible" (Ibid).

Does this correspond to the reality of brain functioning? There are many studies that confirm that neural activity has a high order of timing. Here we will give just one example. A group of researchers made simultaneous recordings in the primary visual cortex of cats with up to 32 electrodes and extracted the preferred temporal sequence of spiking from pairwise cross-correlations of neurons. They made important observations: "(1) Groups of synchronized neurons form firing sequences, each neuron having its own preferred firing time relative to all other neurons in the group. (2) These sequences last shorter than a single cycle of beta/gamma oscillation. (3) Firing sequences are not fixed ("hardwired") but change as a function of stimulus properties. (4) The changes are systematic and permit inferences on stimulus properties. (5) The degree to which firing sequences replicate across repeated stimulus presentations allows for reliable coding" (Havenith et al., 2011).

Izhikevich and colleagues made a simulation of a network of 100,000 cortical neurons with different spike time delays depending on the kinetics of AMPA, NMDA and GABA receptors and the dynamics of long-term and short-term synaptic plasticity. "The network contained large polychronous groups capable of recognizing and classifying quite complicated spatio-temporal patterns ... An unexpected result is that the number of co-existing polychronous groups could be far greater than the number of neurons in the network, sometimes even greater than the number of synapses. That is, each neuron was part of many groups, firing with one group at one time and with another group at another time. This is the main result of the present paper" (Ibid).

The next step was to make a large-scale model. The authors simulated the spatial characteristics of the mammalian thalamocortical network and the dynamic properties of 22 types of neurons. The model included one million neurons and half a billion synapses with appropriate receptor kinetics, short-term and long-term synaptic plasticity. Here is how they describe the result: "The model exhibits behavioral regimes of normal brain activity that were not explicitly built-in but emerged spontaneously as the result of interactions among anatomical and dynamic processes ... The behavior of the model was extremely sensitive to contributions of individual spikes: adding or removing one spike of one neuron completely changed the state of the entire cortex in <0.5 s ... The model spontaneously generated rhythms and propagating waves that had frequency distributions, spatial extents, and propagation velocities similar to those observed in mammalian *in vivo* recordings" (Izhikevich, Edelman, 2008).

Let's look at the results of the experiments from the Symphonic Neural Code (SNC) hypothesis perspective. It predicts that each note played by a neuron has its meaning and significance that is determined by the spatio-temporal parameters of the action potential. The music of the Mind is extremely sensitive to the contributions of individual notes. Depending on the order and precise timing, the notes are rhythmically organized into a meaningful pattern. This rhythmic structure is part of the code and changes as a function of the encoded signal parameters. In addition to the information contained in the waveform of the action potential, the sequence provides further informational density to the code. This code is fast enough to contain meaningful patterns within one cycle of even high-frequency neural oscillation. The result that the number of co-existing "polychronous groups" can be greater than the number of members of the ensemble is expected within SNC. Each neuron can participate in various waves of normal brain activity generating polyphony and polyrhythm of the music of the Mind based on harmonic frequency ratios and phase coupling. Our life depends upon the harmony of this music.

CHAPTER 8

BRAIN MUSIC NOTATION

In a forest of facts or an ocean of thought, one can equally get lost without theories and concepts.

Dmitri Mendeleev

If we take the Symphonic Neural Code (SNC) hypothesis as a working one, then we have no choice but to study the musical notation of the neural network to understand the meaning (internal characteristics) of each note, its place in the general melodic and harmonic structure (population code), rhythmic structure (temporal code), and tempo variations (firing rate). Each parameter will find its place in the musical score of the symphony of the Mind.

In this sense, TTT does not deny previous models but just avoids their limitations and aspects that contradict the physics, physiology and technology of the process. In this study, we have seen how contradictions "magically" disappear, and the data fit into the new concept. This is the main sign that such a theory is a higher-level model since it explains both what has already been explained and what has not yet been explained. It contains more "pieces" of the total data volume.

When trying to assemble a jigsaw puzzle, there are various options. First one: we have no model as an assembly guide. We collect the pieces and try to sort them out hoping that it will work out somehow without a model. Even when it comes to puzzles consisting of a small number of elements, this approach is ineffective, and in the presence of a huge number, it is practically a dead-end. Second option: we have a model, but it is too generalized. The task becomes hopeless if the picture is complex with many details on the pieces themselves and a lot of pieces. As a result, pieces do not fit, whichever way you turn them. The third option: we have a model which is detailed in some places, but there are internal contradictions and gaps in it. It is a puzzle in itself. The task is also hopeless. Pieces will seem to fall into place but then turn out to be wrong. So, it looks like all three options are hopeless.

What should we do? Give up this business and return to mysticism? Should we say that it was a mystery and will remain this way, that it's not our mind's business to understand itself? Such an approach exists, and everyone has the right to such a choice. Anyone who is not satisfied with it will have to look for a fourth option.

Mathematician Ian Stewart said: "If our brains were simple enough for us to understand them, we'd be so simple that we couldn't" (Cohen, Stewart, 1994). There is a pessimistic interpretation: the brain is too complex for it to understand itself. Let's try to look from an optimistic perspective: the brain is hard to understand, but it is complex enough to understand itself. It is up to this task. To do this, it needs to do what it does concerning all the phenomena of this world: create an adequate model.

To effectively assemble a puzzle, we need a holistic, coherent, consistent, at the same time generalized and detailed model, tested for compliance with the task as a whole and with the separate pieces. Unfortunately, we do not have a ready-made ideal model in stock. We have to simultaneously collect the "pieces," twist them this way and that, and "draw" a model that will approach such requirements.

All the previous three approaches are three roads leading to the fourth. And we follow them with this or that success, unless, of course, we give up and return to the very beginning of the mystical awe. Some make pieces of data, some build models, and others do both. Although often in the process, we come across situations when the elements do not want to fit properly, and the temptation comes to somehow correct data or even throw it into the trash and forget, we cannot get away from the puzzle of our own Mind. As we have seen more than once, scientists, having spent their entire lives solving the riddle, often write in their books: this is the business of the next generations. But, on the other hand, there is always a question: who, if not us?

After a lyrical deviation, let's go back to the details that should fill any hypothesis. Again, we are back to the notes of the Mind. Why so much complexity? Why should every note be full of meaning when you can compose code from simple and meaningless elements? Why can the action potentials of neurons not be identical discrete spikes? They cannot be so for a simple reason: neurons are oscillators with specific parameters. They are physical systems and not ideal identical virtual soldiers with the same virtual shots. Maybe the oscillatory nature of cells and the variety of parameters is not a disadvantage but an advantage?

Why do many standard neuroscience theories try to represent neural code as a collection of identical digits? Of course, it's easier to count that way. But there is one more critical point: if the model assumes ideal units of spikes, then there is no need to explain the mechanism of interaction of all these various multidimensional oscillators with many degrees of freedom. We put them into hypothetical chains and let them shoot together or alternately, but strictly with the same spikes.

The only mechanism is the direct connection in the chain (synapse). It remains to express this with a mathematical model, which will take into account either simultaneity or the sequence of spikes and the strength of the connection in the chain (synaptic weight), which will determine whether neurons can fire together or not. And there will be many beautiful diagrams and converging formulas. We can even simulate such a network on a computer and create such a linear artificial neural network.

But real neurons will continue to be who they are: living and multidimensional nonlinear open oscillatory systems that can receive and emit complex and multidimensional signals. Why are they like this? They were born this way. Not because they are so unique, but all living cells are like that. Multicellular systems have adapted some of their cells for the needs of internal communication.

Second, the information density of the neural code, even in a short interval, must be large. This is not just a task for the information network engineer but the task of the organism's survival. We have repeatedly noted an empirical fact: the speed of the system is so high that it is at the limit of the speed of each neuron, within the time frame of several action potentials, within one or two bars of the music of the Mind. This can be compared to the beginning of Ludwig Beethoven's Fifth Symphony:

Several notes begin the symphony and form the basis of all subsequent development. Even those who are not familiar with classical music in detail know these sounds. "This is fate knocking at the door," said Beethoven (or his secretary, but that doesn't matter).

This motif (not even a melodic line, it is so short) generalizes the theme of fate, expressed by the whole symphony: the opposition of the motif of the formidable destined fate to the strong-willed and active nature of life. Everything is here at once: the repetition of a sound as a knock on the door and a forward movement to the first strong beat of the second bar. It is both an epigraph and an architectonic link in the entire complex structure of the symphony, repeated in one form or another in all its parts.

The construction of a massive number of meanings is possible even based on a binary code. In a sense, the music code is also binary: sound/pause. But each sound and pause has its internal characteristics since this is a continuous physical process and not a discrete ideal one or zero.

In a binary code, both parts have a meaning relative to each other: the "zebra" is not black with white stripes and not white with black; it is black-and-white/white-and-black. In a musical code, both the pattern and each symbol of the code carry meaning. Therefore, it is so informationally rich even within one measure and several notes.

Let's elaborate on the Symphonic Neural Code idea.

Hypothesis:

The neural code is physically and semantically analogous to the musical one. The physics of the process is based on continuous oscillatory and wave phenomena. Meaning is embedded in each element of the code and combinations and sequences of elements. Thus, a potentially infinite set of meanings can be created from a limited set of notes. Each note (action potential) is simple enough for the code to be effective. But it also is complex enough to contain information as specific amplitude-frequency parameters and phase portrait.

The transition from the action potential (note) to resting potential (pause) as different phases of the same oscillatory process is conditional and dynamic, and there is no rigid threshold. Thus, a neuron is not a digital computational unit but a hybrid analog-digital device. From the physical point of view, the dynamics of the phase portrait and phase transitions of the neuron as an oscillator set its role in the general ensemble based on synchronization as frequency and phase coupling. From the code's semantics perspective, the neuron's impulse response parameters determine its computational characteristics as an encoder/decoder of signals.

What are the tasks of neurons when processing signals from the external and internal environment?

First, neurons must create information-efficient code.

Second, neurons must create an energetically efficient sparse code (sparseness means a small number of elements in a fast time window).

How can a complex representation be created meeting the above requirements for the operation of neurons? This question has puzzled researchers for years. There is a way out of such a conceptual impasse. Living systems used it to get out of the technological trap of seemingly incompatible code requirements: sparseness and saturation. The system must know the set of probable values of the incoming signal. For this, it has a model that includes the necessary and sufficient set of representations. The receiving decoder must know the statistical structure of the stream. In any code, elements are interdependent, and specific combinations create meaning. Knowing these combinations, the decoder can grasp the meanings on the fly.

The paradox is that with the expansion of the time window and the number of spikes in it, the information saturation by one spike and per unit time begins to fall and tends to zero. In this situation, the system simply cannot wait until a large number of spikes have accumulated to calculate the average speed: it does not have time for this, and the process itself does not make sense since the information entropy (uncertainty) begins to grow. From whatever point of view we look at, the system needs an entirely different code in which every spike, and every pause, and spike/pause pattern matters. Neural code just has to combine sparseness and density. In this situation, both the encoding and decoding parts of the system must work not only efficiently but very precisely coordinated in time up to the resolution level of one spike. To solve these fundamental problems, the system uses the oscillatory nature of the operation of each neuron, the synchronization mechanism, and the accumulation of significant representations of already processed signals to compare the incoming stream with the model.

Similarly, in music: each note contains information; for it to carry meaning in the general flow of music, the parameters of sound vibrations must be inscribed in the melodic, harmonic and rhythmic structure of the entire piece; a musician or listener must have an accumulated representation base to reproduce and perceive information; the encoder (musician) can make up a potentially infinite-dimensional information stream from a minimal set of basic code elements; the decoder (listener) can perceive even new variations very quickly due to the existing model; all variations are possible due to the flexibility of the code as a combination of the elements and the flexibility of the frequency and phase locking mechanism.

No average rate code can provide either information richness in the required time frame or flexibility. It is significant that experiments show how different activity patterns can produce the same number of spikes (average speed). How can a decoder navigate such a code if it just counts the notes and does not pay attention to either the pattern or the notes themselves?

At the beginning of the Fifth Symphony, the "knock of fate" pattern is formed by only two notes. Their number undoubtedly matters but as an increase in the overall information saturation of the beat. But one or two notes can completely change the semantic message. Each signal of the external and internal environment can be called the knock of fate at the door of consciousness. The brain has to encode and decode it as accurately, quickly and efficiently as possible. The precise placement of notes in time and the detailing of the parameters of each note can create meaning even within one measure. From the point of view of information and analysis of the fundamental components of the code, the number of spikes is a component of a simple level, and the parameters of the spikes themselves and their patterns are high-level components. They give the code a high degree of discriminative and generative power, i.e., increase its general informational level. But this also creates enormous difficulties when trying to decipher this code.

As the authors of the book "Spikes: exploring the neural code" wrote: "Once we consider the possibility that arrival time of each spike carries information, the number of dimensions of our description increases dramatically ... With an average number of spikes $\tilde{r} = 30s^{-1}$ and $T \sim 300$ ms, a time resolution $\Delta t \sim 5$ ms gives $2^s \sim 10^{11}$. Thus the number of possible spike trains is of the same order as the (1995) United States government budget measured in dollars. No experiment will ever sample even a tiny fraction of these possibilities; no experimental animal (and few experimenters!) will experience even one-tenth of this number of 300 ms intervals in a lifetime. Clearly one cannot jump from counting spikes to a "complete" analysis of timing codes without some hint about how to control this explosion of possibilities" (Rieke et al., 1999).

Indeed, on the one hand, the transition to the analysis of subtle temporal nuances of neural activity leads to an explosion of possible options, which creates both technical and analytical difficulties. This is, of course, intimidating, and the researchers are not ready to "jump." It is much more comfortable to stay in the usual mainstream looking at the average firing rate. But this counting has been going on for almost a hundred years but did not give anything for deciphering because the code is different. On the other hand, the explosion of analysis possibilities also creates a potential explosion of opportunities to move towards the goal, if it is about deciphering the code and not obtaining grants and titles, which are usually distributed within the framework of established paradigms.

Here we need to take an example from our neurons again. "Over a time window of one second, the neuron is providing a spike which uniquely identifies one signal out of $2^{300} \sim 10^{90}$ possible signals" (Ibid). Such a volume is hard to imagine, but the brain simply has no other choice but to cope with a potentially endless array of signals. The neural code turned out to be not as simple as we thought because the world it encodes is not simple. The brain is complex, but, as we have already noted, with an optimistic look, this means that it can analyze itself.

With all the potentially endless variability of the messages, the building blocks can be quite simple. The music consists of not such a large set of basic elements and patterns, but variations have an astronomical volume. Neural code shouldn't be too complex or too simple. If the code is too simple, its informational richness is low, and the informational entropy (uncertainty) will be too high. If it is too complex, then again, there will be unreasonably high uncertainty. Extremes converge in the same negative result. The golden mean is that the code is both rich and digestible. The system elements also need to understand each other with a sufficient degree of certainty; therefore, the code has specific patterns and a set of basic patterns. There is only one thing left to do: find them.

The beginning can be the formulation of hypotheses and a general concept, and then the setting of tasks for the study of these hypotheses, the creation of measurement technologies and measurement analysis. Much has already been done, many developments have been made in other areas of knowledge related to large data analysis. The most important thing is that the volume does not flood the researcher. This requires a "lifeline" of the conceptual approach.

Of course, not all researchers just average firing rate. Since the emergence of technologies that make it possible to measure the direct activity of neurons in living tissue using electrodes, they began to try to identify patterns in the operation of a neural network depending on different incoming signals.

Here is an example of one experiment in which the approach differed from traditional methods. The authors did not aim to decipher the code and did not highlight the details of the internal dynamics of action potentials. They tried to answer a question: what should the input signal on the electrode be for the response activity of neurons to have a pattern and not look like random noise?

The activity of neurons is aperiodic since they are complex oscillators that work as nonlinear filters. In a musical analogy, we can put it like this: neurons are not ticking metronomes, but musicians playing complex rhythmic and melodic parts. In terms of chaos theory, neurons are chaotic systems. That is why the authors' article was titled "Controlling chaos in the brain" (Schiff et al., 1994). It sounds as if the brain is in a mess, and the experimenters want to put things in order. Everything is exactly the opposite: the brain has a nonlinear order of complex patterns (chaos in this sense), and researchers are only trying to make their messages (electrode signals) less messy and more understandable for the receiving system. Imagine inmates in different prison cells who want to communicate by knocking on the wall but do not know any common code. Through trial and error, they can send out specific patterns of knocking and try to hear the patterns in response and then begin to create a common language.

Or another comparison: neuroscientists are like space explorers, sending signals hoping that intelligent beings will accept it and respond. They usually send structured patterns. By the way, very often, these are pieces of music. But they are not sending a metronome beat, although this is a very structured pattern. The paradox is that such a clear periodic signal practically does not carry any information. Only a complex aperiodic signal generates information. And we hope that the receiving side can decode such a signal if it is intelligent, i.e., it is a teleological signal transducer.

When trying to start a dialogue in an unfamiliar language, you need to proceed with simple things. If we are trying to talk with the cosmos of the brain, then we need to start with the most basic patterns, even with a periodic signal. Although neurons play complex parts, they are able to "hear" the metronome. Periodic basic pulsations underlie any music, including the music of the Mind.

We do not know the neural code and cannot send any complex meaningful message. It remains to make our signal more or less periodic and see if the response will have a structure. If we "knock" to the neuron, it should "knock" back to us. After all, it is the intelligent being. The authors described the task by chaos theory concepts: chaotic systems are sensitive to small perturbations; if the neural network is a chaotic complex system, it is necessary to find such a mode of external influence that will cause a specific response. It will be expressed in a change in the phase space trajectories, i.e., show the dynamics of parameters with identifiable patterns.

As the authors wrote: "Of one observes the timing of events from a chaotic physical system, those events are aperiodic. The timing of events evolves from one unstable periodicity to another. Furthermore, the approach to these unstable periodicities shows recurring patterns which can be quantitatively understood by examining the relationship between the timing of sequential events" (Ibid).

In musical analogy, we can say that for all the complexity and external aperiodicity, the music of a complex system has a rhythmic structure (durations and sequences) that can be calculated. You need to highlight some of the brightest patterns and build a timeline relative to this reference mark. For example, in music, these are usually repetitive strong beats. If we listen to the music of sounds, our brains automatically calculate this rhythm, even if we are not counting on a cognitive level.

What should neuroscientists do when they look (although it would probably be more effective to listen) at the signals of neurons? It is necessary to apply an appropriate graphical visualization method and analysis algorithm. "This can be visualized using a plot which is a type of a return map. Such a map plots the present interval between events versus the previous interval. Periodicities are revealed on such a plot as intersections with the line of identity. In a chaotic system, these intersections are known as unstable fixed points. These unstable fixed points are deterministically approached from a direction called the stable direction or manifold and exponentially diverge from these points along the unstable direction or manifold. This type of local geometry has the shape of a saddle. Chaos control consists of the identification and characterization of these saddles, followed by a control intervention which exploits the local geometry of the saddle to increase the periodic behavior of the system. We demonstrate three separate approaches to control: periodic pacing, an implementation of chaos control theory, and the inverse of chaos control, which we term anticontrol" (Ibid).

To get through the terminology of dynamical systems theory, we need to think again about music. If we imagine the "line of identity" as a periodic basic pulsation embedded in any musical fabric, then the unstable fixed points (UFPs) will be beats of a musical measure. They are always present, even if there is no note on the beat itself. Especially strong beats will be such anchoring points. The entire rhythmic structure of all notes will be "unstably fixed" around the beats. They can fall into them, they can be in between, but they are always in a clear relationship with the basic structure of the measure, depending on the duration and position of a given note. Otherwise, in the absence of such a connection with the "line of identity," everything will decay into noise without a pattern (chaos in the ordinary sense).

If we represent such a distribution graphically, the basic pulse will indeed be a certain line, and all notes will be dots around it. If this is music, then the points will be in a compact space around this line and form a kind of saddles, i.e., converge and diverge but remain locked to that axis. If it is noise, then the points will scatter across the "field" like frightened rabbits. "Controlling chaos" in this case will be intervening in such a run and bringing the system into a structured form around the line. It can be done by turning on the metronome (usually done during training or rehearsals), looking at the conductor, listening to the rhythmic bass line, and not forgetting about the inner feeling of pulsation (usually done during the actual performance). This is "chaos control" in music.

The method of analysis used by the authors is called Poincaré maps or return maps. It shows the points at which the continual trajectory of the development of the system's phase space "visits" the same region. Any physical process can have both vivid periodic behavior and hidden irregular cyclicities (chaotic attractor). If the system is stable and coherent, then the states do not run off exponentially, and after some time, the divergences return to the attractor. This is a fundamental property of nonlinear chaotic systems, for which there is even a special Poincaré return theorem.

The graphical representation of a continuous process in the form of discrete points around a specific region of attraction simply facilitates the analysis of the system: instead of continual trajectories that can intersect and scatter, creating complex "tangles of threads," the analyst gets discrete return points and their distribution. These points indicate intersections, where the system repeats the pattern of the dynamics of its parameters.

So, the authors applied this method of analyzing neuron signals to understand whether they have any patterns and whether they can be related to the patterns of an external signal. They called this "chaos control." They called experiments with missing patterns in the incoming signal "anticontrol". In our analogy, they either turned on the metronome, acted as the conductor or rhythm section of the ensemble, or confused the musicians with various noises. The task of the experiment can be formulated as follows: to determine whether neurons are playing music or making random noise.

The researchers took live sections of rat hippocampus tissue, placed it in artificial cerebrospinal fluid, applied signals with electrodes, and measured neuronal signals. For a periodic signal, they used not some arbitrary external pulsation, but analyzed the signals of neurons, identified UFPs, the speed of approach and departure from them, and used the same pulse interval as the calculated UFPs in subsequent experiments. It was an attempt to define the beats and measure of this music. For anticontrol, an interval was chosen that placed the next point on a line that was entirely outside the set. This anticontrol technique effectively deviated the state of the system from a stable direction and, therefore, from UFP (Ibid). They made 91 attempts on 22 tissue sections from 9 rats. Here is the return map:

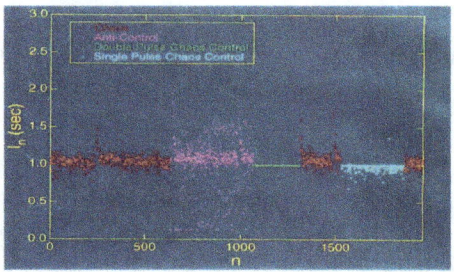

It clearly shows that the neurons before the impact (red part) play their music compactly around the baseline. After anticontrol (deliberate distraction from the rhythm), they begin to leave the area of coherence (pink part). Then chaos control is switched on with an amplified double periodic signal, and everything converges into one line. Then there is "freedom of flight" (red part), and at the end, control with a single periodic signal, where the line is still clear but with some deviations (blue part). It is literally obvious how neurons-musicians can play their coherent music, listen to the metronome, listen to the rhythm section or follow the conductor. But in case of strong interference, they begin to get confused and at times slide into noise.

This is how it looked if the neuron signals were placed on the time axis (similar to a musical staff but without the pitch axis):

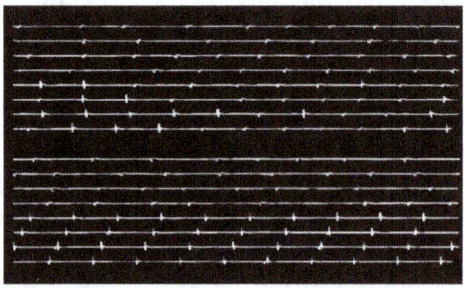

The authors do not consider action potentials as continuous processes with internal parameters but only place neuron signals compressed in time and reduced to discrete spikes on the line trying to identify the presence of patterns. The notes are reduced to identical points without any differentiation, but this is enough to reveal the rhythmic structure of the sequences. Here you can see how, when the incoming control signal is perceived, a reasonably apparent periodicity is established. Moreover, as the control signal becomes more stable, the responses become almost perfectly periodic. In response to the knock-knock on one side of the "wall," a knock-knock appeared on the other side — it means there is someone intelligent there.

If you look closely at the recording in the lower part, you can see how immediately after the beginning of the "controlling" periodic signal, a systematic response is established, but then bursts occur a little earlier of incoming pulse. The experimenters do not comment on this in any way. By the way, they generally do not draw any theoretical conclusions from the experiment, except for the practical one that it is possible to control the activity of neurons and thus produce a therapeutic effect, for example, in epilepsy. But their experiment says not only that external signals can influence neurons. This is nothing new. The results are much deeper than the authors' modest conclusions.

Let's go back to our musical analogy. It seems that neurons are starting to syncopate: they shift their rhythmic emphasis relative to the metric alternation of beats. They remain in a precise rhythm, but shifted relative to the metronome. They respond to knock-knock with their knock-knock moved in time relative to the received signal. Is it logical? Sure. Our neurons (sorry, rat neurons in this case) are quite logical, unlike some theories about them. If they "knock" at the same time as the incoming message, then what kind of dialogue can happen? If you speak at the same time (fire together), there will be no conversation. A phase shift is necessary, but with the preservation of stable coupling as a manifestation of interaction and synchronization.

A simple idea: if the signal supplied by the electrodes has some pattern, the receiving neurons perceive it as a signal with meaning, process it, and give out their pattern-meaning. And if the incoming signal is a simple periodic knock-knock, then the neurons respond with the same simple knock-knock, but in their own time. If a random signal is sent to them (anticontrol), the response becomes aperiodic but retains the structure around the identity line that existed before the signal, although a scattering region appears. This is how neurons "tell" the experimenters: although you send us nonsense, we are still reasonable and keep the coherence of our meanings.

At first glance, it seems strange that the response begins to outpace the message. The authors called this "escape behavior," but from the point of view of the PAAL model, everything is precisely the opposite: it is "running to" behavior. Let's think about the course of the experiment. The researchers cut out the brain tissue, ripped out part of the neural network from the established connections but kept the neurons alive by placing them in saline. They began the learning period when the activity of neurons was measured and influenced by different impulses, and then the control phase of the experiment started. These neurons have impulse response settings as their share in the reality model of the system they used to be a part of. But they also have the ability to tune in to changes in the environment and are not passive receivers but active creators of meanings (representations).

The researchers used slices of the hippocampus. According to the hypothesis within the TTT framework, this area of the brain is populated by specialists in creating new patterns during the comparison of the projected representations with the introjected environmental signals, the correction of old representations, and the renewal of the reality model. They are at the crossroads; they are at the center of the process.

Is it surprising that they first manifest endogenous patterns, then begin to listen sensitively to the incoming signal from the electrodes, create a similar pattern and synchronize with an external signal, and then start to work ahead and show the same pattern but before the incoming signal? Nothing is surprising if we take the hypothesis about the PAAL algorithm as a working one with explanatory and predictive power. In this case, this behavior of neurons becomes predictable and explainable. The experiment demonstrated PAAL in action even in a piece of cut-off tissue: a constant iterative flow of projection-introjection, synchronization with external signals, and synchronization of internal processes with each other. Another piece of the puzzle just fell into its place prepared by the general concept.

During the experiment, neurons "told" the researchers: we have received your message; we have already coded it and understood it; we have incorporated it into our model of reality; we are projecting it ahead of the curve; we are ready for something new. But for a full-fledged dialogue, including for specific therapeutic purposes, such "knocking" is not enough since a transition from a simple periodic signal, as an initial "handshake," to a real and complex code is necessary. This requires further detailing of our messages and analysis of the messages of neurons. Averaging, which continues to be the basis of research in neuroscience, is simply lethal for such a dialogue: it ends before it starts. It is not surprising that decades of this approach did not lead to any dialogue with neurons (their language remained a mystery). Perhaps Izhikevich was right that all this was a waste of time.

The example of this experiment shows: when we pay attention to the rhythmic structure of the spikes (even without detailing the internal dynamics of the spikes themselves), place the notes at least on the time axis (the frequency axis is absent in the authors' analysis) and stop averaging these notes in an attempt to calculate the average speed, then the activity of neurons immediately manifests intelligence (meaningful patterns).

By the way, if the experimenters instead of a visual graphical representation of the results would make an audio one, then it would be no less, if not more convincing. They would hear the neurons gradually synchronize with the external metric pulse. One could dance to this music of the Mind. Imagine a picture: scientists in the laboratory connected the signals of their electrodes and neurons to the converting audio equipment, got a soundtrack, and danced to the experiment results. This is trivial from a technical point of view, but is beyond the conceptual barrier. Many neuroscientists call the brain an orchestra but they still take this as a loose metaphor and do not think in musical terms from a physical and technical perspective. Now imagine that in addition to recording the rhythm scientists would expand the individual characteristics of the notes of the Mind and place them on the staff along axes of space and time. Then we would hear not only the drumbeat of discrete pulses but also melodies and harmonies of the music of neurons. Such a result could claim both the Nobel Prize and the Grammy.

In the above experiment, discrete pulses of electrodes were applied and discrete pulses of neurons were analyzed. Now, what if we apply waves? In general, the idea of stimulating the brain with exogenous electromagnetic waves has a long history in neuroscience. We can say that it started as soon as electromagnetism was discovered and described in physics in the 19th century. But as the authors of one experiment wrote, despite advances in technology and many attempts to influence brain activity by various electromagnetic devices, "there is still a lack of understanding at the basic cellular level of what happens when brain matter is electrically stimulated to promote certain activity patterns or suppress others. This has limited the application of electric stimulation as a basic science tool and, importantly, as a therapeutic intervention for brain disorders" (Lee et al., 2023).

The authors took slices of rat and human brains and applied electric stimulation (ES) using the whole-cell patch-clamp technique. It enables simultaneous extracellular stimulation, intracellular current injection, and extracellular recordings with high definition and close to identified neurons. They delivered sinusoidal ES of varying amplitude and frequency. Their first observation is that at "subthreshold, rest and polarized levels, the cell membrane can follow the applied ES and does so robustly across all cell classes." The second observation is that while neurons "show strong, class-specific spike entrainment to ES, the same neurons exhibit indistinguishable entrainment characteristics to the same sinusoidal ES when not spiking." The authors conclude that class-specific synch is "the result of two effects: first, a non-specific membrane polarization ubiquitously present across cells and classes; second, class-specific excitability and activity properties that allow spikes of similar time scale to the ES to entrain particularly strongly" (Ibid).

The analysis of the frequency structure showed "that distinct cortical classes with vastly different spiking profiles across brain areas and species exhibit ubiquitous and robust coordination to ES in a spectrum of frequencies. This observation suggests that membrane entrainment is a fundamental property ... Yet, since how a cell spikes (from its spike waveform to the excitability profile and its resulting spike rate) is crucially affected by its ionic setup, the distinct ionic setup of each class is implicitly reflected in the properties of ES entrainment" (Ibid).

Let's translate this into the physics of music. All participants of the ensemble have the fundamental property of being able to synch. It does not depend upon their class-specific characteristics. Whether it is a violin or a cello that sound differently due to harmonics in the timbre, they sync with the basic pulse of the symphony. The pulse can be described as a wave with strict periodicity. In the experiment, the sinusoidal ES signal was the exogenous pulse to which neurons "listened" and entrained. They did it robustly across cell classes even during subthreshold and resting state, thus once again showing that their activity is not discrete spikes but waves of continuous oscillations.

In music, the pulse continues even if notes do not sound at all. It is a metric of a temporal stability on which all events hang as on a thread. The fundamental frequency of the pulse is the "clock" of the interacting oscillators that brings them all into a coherent system. Even syncopated rhythms that fall between strong beats of the meter are within this basic pulse. But it is not a set of discrete points in time. It is a wave structure with a stable period. The phases of this wave can be highlighted as the strong and weak beats of the meter. These points appear on the timeline according to the frequency of the wave.

Intervals between pulses of the metrical structure are filled by multiple frequency components (rhythms, notes, melodies, chords) with harmonic phase and frequency relationships both with the pulse and with each other. The same happens in the music of the brain. From a physical perspective, the slices of the neural tissue used in the experiment are networks of oscillators that were stimulated by the exogenous periodical wave and showed not only the entrainment to this wave but the multifrequency pattern in the wide range of cell-specific frequencies that depend upon the characteristics of the cell, including ionic setup. The authors speak about a spike waveform. This is a major step forward from the standard attitude to the activity of neurons as identical shots.

But why do neurons respond so complexly to a simple sinusoidal signal? It seems to be a waste of energy. They do this for a practical reason: they need to encode a set of parameters that will allow to classify the signal. Even a simple periodic signal can have a different waveform. Moreover, real signals are not pure sine waves. When a violin and a cello play the same note C in the same octave the resulting sound is full of harmonics around the fundamental frequency. Our brain encodes this sound as the representation of a violin C and a cello C that differ from each other in timbre. This can only be done by the high variability of the encoders' activity profile.

This variability is a nuisance for the researchers who consider it redundant and just average it out on their whim. But why would nature develop such a variety if all that it takes to encode a complex world is to fire at different average speeds of identical shots? This sounds like a paradox. But the real paradox is that, despite the observed variety of the activity profile of the neurons, there is "a widespread belief that neurons are inherently noisy, and ideas of redundancy and averaging pervade much of the literature" (Rieke et al., 1999). Maybe the point is not that neurons are noisy, but that we do not understand the meaning of their messages. Fortunately for us, neurons and their signals, even at the level of one action potential, are meaningful. This, of course, is good for us as users; otherwise, we would have no consciousness. But it is a complication for us as researchers of our own neurons.

The observed speed and precision of encoding also causes complications for theorists. "Echolocating bats can apparently resolve jitter in the arrival time of their echoes with a precision of 10 nanoseconds. The weakly electric fish are not far behind adjusting their own electrical signals in response to several hundred nanosecond shifts in signals from neighboring fish. These results are surprising because the natural scale of neural activity seems to be in milliseconds, not microseconds, and certainly not nanoseconds" (Ibid). How can neurons that produce a sequence of spikes in milliseconds encode signals on the nanosecond scale? The scale difference is 10^6. This is another paradox for mainstream theories. But the real paradox is that, despite the observed encoding speed, they obstinately talk about slow spike trains and turn a blind eye to fast intra-spike and inter-spike details.

"Given the evidence for performance close to the physical limits, it certainly makes sense to ask which features of neural coding and computation are essential to this remarkable performance. What is the structure of a neural code that allows such high rates of information transmission? ... We would like to have a theory of the computations required to make estimates and decisions at this limiting level of reliability ... Our story began, more or less, with Adrian's discovery that spikes are the units out of which our perceptions must be built. We end with the idea that each of these units makes a definite and measurable contribution to those perceptions. The individual spike, so often averaged in with its neighbors, deserves more respect" (Ibid).

Yes, each note deserves more respect for two reasons: first, without it, there would be no music that we call the Mind; second, without understanding and analyzing all the subtleties of each note, we will not be able to read the notation of the brain. Such a task seems at the moment almost as insurmountably difficult as for a novice musician reading Beethoven's symphonies:

Musicians learn to read the notation step by step and the first step is to know the fundamental elements. We have been ignoring such elements of the brain notation and found ourselves ignorant of the music of the Mind. It is time to get back to basics.

The neural code turned out to be not as simple as we thought because the world it encodes is not simple. The Teleological Transduction Theory prediction which makes the task a little easier is that the brain uses a limited set of standard variations to produce an internal universe of the reality model. Just like twelve notes can produce the musical universe.

REFERENCES

Adrian, ED, Zotterman, Y. (1926). *The impulses produced by sensory nerve endings: Part II: The response of a single end organ.* J Physiol (Lond.). 61: 151–171.

Ajina S, Bridge H. (2016). *Blindsight and Unconscious Vision: What They Teach Us about the Human Visual System.* Neuroscientist. 2016 Oct 23;23(5):529-541. doi: 10.1177/1073858416673817.

Baars, B. J. (1997). *In the Theater of Consciousness.* New York, Oxford University Press.

Balduzzi, D; Tononi, G (2008). *Integrated information in discrete dynamical systems: motivation and theoretical framework.* PLOS Comput Biol. 4 (6): e1000091.

Ballard, D. H., Jehee, J. (2012). *Dynamic coding of signed quantities in cortical feedback circuits.* Frontiers in Psychology, 3(254).

Barabasi, Albert-László, Albert, Réka. (1999). *Emergence of scaling in random networks.* Science 15 Oct 1999: Vol. 286, Issue 5439, pp. 509-512.

Barbour, Julian. (1999). *The End of Time: The Next Revolution in our Understanding of the Universe.* Oxford Univ. Press. ISBN 0-297-81985-2

Barlow, H. B. (1961). *Possible principles underlying the transformation of sensory messages*. Sensory Communication, pp. 217–234.

Bassett, D.S., Meyer-Lindenberg, A., Achard, S., Duke,T., Bullmore, E. (2006) *Adaptive reconfiguration of fractal small-world human brain functional networks.* PNAS December 19, 2006 vol. 103 no. 51

Bastos, A.M., Vezoli, J., and Fries, P. (2015). *Communication through coherence with inter-areal delays.* Current opinion in neurobiology 31, 173-180.

Belluscio, M.A., Mizuseki, K., Schmidt, R., Kempter, R., Buzsaki, G. (2012). *Cross-frequency phase-phase coupling between theta and gamma oscillations in the hippocampus.* J Neurosci 32, 423-435.

Bichot, N.P., Rossi, A.F., and Desimone, R. (2005). *Parallel and serial neural mechanisms for visual search in macaque area V4.* Science 308, 529-534.

Bose, A., Recce, M. (2011). *Phase precession and phase-locking of hippocampal pyramidal cells.* Hippocampus. 11(3): 204-15.

Bosman, C.A., Schoffelen, J.M., Brunet, N., Oostenveld, R., Bastos, A.M., Womelsdorf, T., Rubehn, B., Stieglitz, T., De Weerd, P., and Fries, P. (2012). *Attentional stimulus selection through selective synchronization between monkey visual areas.* Neuron 75, 875-888.

Bragin, A., Jando, G., Nadasdy, Z., Hetke, J., Wise, K., Buzsaki, G. (1995). *Gamma (40-100 Hz) oscillation in the hippocampus of the behaving rat.* J Neurosci 15, 47-60.

Brincat, S.L., Miller, E.K. (2015). *Frequency-specific hippocampal-prefrontal interactions during associative learning.* Nat. Neurosci. 18, 576–581.

Brzezicka A, Kaminski M, Kaminski J, Blinowska K. (2011). *Information transfer during a transitive reasoning task.* Brain Topogr. 2011; 24:1–8.

Buschman TJ, Miller EK. (2007). *Top-down versus bottom-up control of attention in the prefrontal and posterior parietal cortices.* Science. 2007; 315:1860–1862.

Buzsáki, G. (2006). *Rhythms of the brain.* Oxford: Oxford UP.

Buzsáki, G., Watson, B. (2012). *Brain rhythms and neural syntax: implications for efficient coding of cognitive content and neuropsychiatric disease.* Dialogues in Clinical Neuroscience.Vol 14, No. 4, 2012 .

Buzsáki, György, Anastassiou, Costas A., Koch, Christof (2012). *The origin of extracellular fields and currents — EEG, ECoG, LFP and spikes.* Nature Reviews. Neuroscience Volume 13 June 2012.

Buzsáki, G., Logothetis, N., Singer, W. (2013). *Scaling Brain Size, Keeping Timing: Evolutionary Preservation of Brain Rhythms.* Neuron 80, October 30, 2013.

Cajal, S. (1911). *Histologie du Systeme Nerveux de l'Homme et des Vertebres.* Paris: Maloine.

Canolty, R.T., Edwards, E., Dalal, S.S., Soltani, M., Nagarajan, S.S., Kirsch, H.E., Berger, M.S., Barbaro, N.M., and Knight, R.T. (2006). *High gamma power is phase-locked to theta oscillations in human neocortex.* Science 313, 1626–1628.

Canolty, R.T., Knight, R.T. (2010). *The functional role of cross-frequency coupling.* Trends Cogn Sci 14, 506-515.

Cavanagh JF, Cohen MX, Allen JJ. (2009). *Prelude to and resolution of an error: EEG phase synchrony reveals cognitive control dynamics during action monitoring.* J Neurosci. 2009; 29:98–105.

Cohen, Jack, Stewart, Ian (1994). *The Collapse of Chaos: discovering simplicity in a complex world* . Penguin Books, 1994, ISBN 978-0-14-029125-4.

Cohen MX, Cavanagh JF. (2011). *Single-trial regression elucidates the role of prefrontal theta oscillations in response conflict.* Front Psychol. 2011; 2:30.

Colgin, L.L., Denninger, T., Fyhn, M., Hafting, T., Bonnevie, T., Jensen, O., Moser, M.B., Moser, E.I. (2009). *Frequency of gamma oscillations routes flow of information in the hippocampus.* Nature 462, 353-357.

Deco, Gustavo Cabral, Joana, Woolrich, Mark W., Stevner, Angus B.A., van Hartevelt, Tim J., Kringelbach, Morten L. (2017). *Single or multiple frequency generators in on-going brain activity: A mechanistic whole¬brain model of empirical MEG data.* Neuroimage. 2017 May 15; 152: 538–550.

Dehaene, S.; Sergent, C.; Changeux, J.-P. (2003). *A neuronal network model linking subjective reports and objective physiological data during conscious perception.* Proceedings of the National Academy of Sciences. 100 (14): 8520–8525.

Dehaene S, Changeux JP, Naccache L, Sackur J, Sergent C. (2006). *Conscious, preconscious, and subliminal processing: a testable taxonomy.* Trends Cogn Sci. 2006 May;10(5):204-11.

Dehaene S, Charles L, King JR, Marti S. (2014). *Toward a computational theory of conscious processing.* Curr Opin Neurobiol. 2014 Apr; 25:76-84.

Dennett, Daniel (1981). *Brainstorms: Philosophical Essays on Mind and Psychology.* MIT Press. ISBN 0262540371.

Dong, Y.; Mihalas, S.; Qiu, F.; von der Heydt, R. & Niebur, E. (2008). *Synchrony and the binding problem in macaque visual cortex.* Journal of Vision, 8 (7): 1–16, doi:10.1167/8.7.30, PMC 2647779, PMID 19146262

Donner, T.H., Siegel, M., Fries, P., Engel, A.K. (2009). *Buildup of choice-predictive activity in human motor cortex during perceptual decision making.* Curr Biol 19, 1581-1585.

Engel AK, König P, Kreiter AK, Singer W. (1991a). *Interhemispheric synchronization of oscillatory neuronal responses in cat visual cortex.* Science. 1991; 252:1177–1179.

Engel AK, Kreiter AK, König P, Singer W. (1991b). *Synchronization of oscillatory neuronal responses between striate and extrastriate visual cortical areas of the cat.* Proc Natl Acad Sci U S A. 1991; 88:6048– 6052.

Engel, Andreas K, Fries, Pascal. (2010). *Beta-band oscillations — signalling the status quo?* Current Opinion in Neurobiology 20:156–165.

Ermentrout, G. B., Kopell, N. (2000). *Mechanisms of Phase-Locking and Frequency Control in Pairs of coupled Neural Oscillators.* Handbook of Dynamical Systems.

Farrior, C.E. et al. (2016). *Dominance of the suppressed: Power-law size structure in tropical forests.* Science Jan 8, 2016 p 155.

Feldman J. (2013). *The neural binding problem(s).* Cogn Neurodyn. 2013 Feb;7(1):1-11.

Fell, J., Klaver, P., Lehnertz, K., Grunwald, T., Schaller, C., Elger, C.E., and Fernandez, G. (2001). *Human memory formation is accompanied by rhinal-hippocampal coupling and decoupling.* Nat. Neurosci. 4, 1259–1264.

Fingelkurts AnA, Fingelkurts AlA, Krause CM, Kaplan AYa, Borisov SV, Sams M. (2003a) *Structural (operational) synchrony of EEG alpha activity during an auditory memory task.* NeuroImage 2003; 20:529–42.

Fingelkurts AnA, Fingelkurts AlA, Krause CM, Möttönen R, Sams M. (2003b) *Cortical operational synchrony during audio-visual speech integration.* Brain and Language 2003; 85:297–312.

Fingelkurts AnA, Fingelkurts AlA, Kallio-Tamminen T. (2020). *Selfhood triumvirate: From phenomenology to brain activity and back again.* Consciousness and Cognition, Volume 86, 2020, 103031, ISSN 1053-8100.

Fingelkurts AnA, Fingelkurts AlA. (2022). *Depersonalization Puzzle: A New View from the Neurophenomenological Selfhood Perspective.* 1. 181-202. 10.5281/zenodo.7253994.

Freeman, W. (1962). *Linear approximation of prepyriform evoked potential in cats.* Exp. Neurol. 5, 477–499.

Freeman, W. J. (1972). *Waves, Pulses, and the Theory of Neural Masses.* Progress in Theoretical Biology, Vol. 2, 1972, Academic Press, Inc., New York and London.

Freeman, W. J. (1978). *Models of the dynamics of neural populations.* Electroencephalogr. Clin. Neurophysiol. Suppl. 34, 9–18.

Freeman, W. J. (1981). *A physiological hypothesis of perception.* Perspectives in Biology and Medicine 560-592.

Freeman, W. J. (1990). *On the Fallacy of Assigning an Origin to Consciousness.* Journal Article e–Reprint. Reprinted from Chapter 2 in: John ER (ed.) Machinery of the Mind Cambridge MA, Birkhaeuser Boston. pp. 14-26.

Freeman, W. J. (1991a). *The Physiology of Perception.* February 1991 Scientific American, Vol 264, (2) Pgs. 78-85.

Freeman, W. J. (1991b). *A novel pathway into brain dynamics (Unpublished Introduction).* Journal article e-reprint 23 January 1991.

Freeman, W. J. (1992). *Tutorial on neurobiology: from single neurons to brain chaos.* International Journal of Bifurcation and Chaos 2: 451-482. 1992

Freeman, W. J. (1999). *Consciousness, Intentionality, and Causality.* Journal of Consciousness. Studies 6 Nov/Dec 1999: 143-172.

Freeman, W. J. (2004). *How and Why Brains Create Meaning from Sensory Information.* International Journal of Bifurcation and Chaos, 14(2).

Freeman, W. J. (2008). *Nonlinear Brain Dynamics and Intention According to Aquinas.* Mind & Matter Vol. 6(2), pp. 207–234.

Freeman, W.J., Skarda, C.A. (1990a). *Representations: Who Needs Them?* J.L. McGaugh, N. Weinberger & G. Lynch eds., Brain Organization and Memory Cells, Systems & Circuits, New York: Oxford University Press, 375-380.

Freeman, W.J., Skarda, C.A. (1990b). *Chaos and the new science of the brain.* Concepts in Neuroscience, Vol. 1, No. 2 (1990) 275–285 World Scientific Publishing Company.

Freeman, Walter J., Skarda, Christine A. (1991). *Mind/Brain Science: Neuroscience on Philosophy of Mind.* in E. Lepore & R. Van Gulick eds., John Searle and his Critics, Oxford: Blackwell, 115-127.

French, K. (2012). *The Hidden Geometry of Life: The Science and Spirituality of Nature.* Watkins Publishing.

Fries, P. (2005). *A mechanism for cognitive dynamics: neuronal communication through neuronal coherence.* Trends Cogn. Sci. 9, 474–480.

Fries, P. (2015). *Rhythms for Cognition: Communication through Coherence.* Neuron, 88(1), 220–235.

Friese, U., Koster, M., Hassler, U., Martens, U., Trujillo-Barreto, N., and Gruber, T. (2012). *Successful memory encoding is associated with increased cross-frequency coupling between frontal theta and posterior gamma oscillations in human scalp-recorded EEG.* Neuroimage 66C, 642–647.

Fuchs, E.C., Zivkovic, A.R., Cunningham, M.O., Middleton, S., Lebeau, F.E., Bannerman, D.M., Rozov, A., Whittington, M.A., Traub, R.D., Rawlins, J.N., and Monyer, H. (2007). *Recruitment of parvalbumin-positive interneurons determines hippocampal function and associated behavior.* Neuron 53, 591–604.

Fyhn, M, Molden, S, Witter, M, Moser, E, Moser, M. (2004). *Spatial representation in the entorhinal cortex.* Science 305 (5688): 125 8–64.

Glomb, K., Kringelbach, M., Deco, G., Hagmann, P., Pearson J., Atasoy S. (2021). *Functional harmonics reveal multi-dimensional basis functions underlying cortical organization.* Cell Reports, Volume 36, Issue 8, 2021.

Gold, Carl, Henze, Darrell A., Koch, Christof, Buzsáki, György (2006). *On the Origin of the Extracellular Action Potential Waveform: A Modeling Study.* J Neurophysiol 95: 3113–3128.

Granovetter, M. S. (1973). *The strength of weak ties.* American Journal of Psychology, 78 (6), pp. 1360—1380.

Gray, C. M., Singer, W. (1989). *Stimulus-specific neuronal oscillations in orientation columns of cat.* Proc Natl Acad Sci U S A. 1989 Mar;86(5):1698-702.

Gregoriou GG, Gotts SJ, Zhou H, Desimone R. (2009). *High-frequency, long-range coupling between prefrontal and visual cortex during attention.* Science. 2009; 324:1207–1210

Grothe, I., Neitzel, S.D., Mandon, S., and Kreiter, A.K. (2012). *Switching neuronal inputs by differential modulations of gamma-band phase-coherence.* The Journal of neuroscience : the official journal of the Society for Neuroscience 32, 16172-16180.

Haegens, S., Osipova, D., Oostenveld, R., Jensen, O. (2010). *Somatosensory working memory performance in humans depends on both engagement and disengagement of regions in a distributed network.* Human brain mapping 31, 26-35.

Haegens, S., Nácher, V., Hernández, A., Luna, R., Jensen, O., Romo, R. (2011). *Beta oscillations in the monkey sensorimotor network reflect somatosensory decision making.* Proceedings of the National Academy of Sciences 108, 10708.

Hafting T, Fyhn M, Bonnevie T, Moser MB, Moser EI. (2008). *Hippocampus-independent phase precession in entorhinal grid cells.* Nature. 2008; 453:1248–1252.

Handel, B., Haarmeier, T. (2009). *Cross-frequency coupling of brain oscillations indicates the success in visual motion discrimination.* NEUROIMAGE 45, 1040-1046.

Harris, Kenneth D., Henze, Darrell A., Hirase, Hajime, Leinekugel, Xavier, Dragoi, George, Czurko, Andras, Buzsáki, Gyorgy (2002). *Spike train dynamics predicts theta-related phase precession in hippocampal pyramidal cells.* Nature. Vol. 417. 13 June 2002.

Havenith MN, Yu S, Biederlack J, Chen NH, Singer W, Nikolić D. (2011) *Synchrony makes neurons fire in sequence, and stimulus properties determine who is ahead.* J Neurosci. 2011 Jun 8;31(23):8570-84.

Hawkins, Jeff, Blakeslee, Sandra (2005). *On intelligence.* Times Books, Henry Holt and Co.

Hawkins, Jeff (2021) *A Thousand Brains: A New Theory of Intelligence.* March 2nd 2021. Basic Books.

Heinrichs-Graham, Elizabeth, Arpin, David J., Wilson, Tony W. (2016). *Cue-related temporal factors modulate movement-related beta oscillatory activity in the human motor circuit.* J. Cogn. Neuroscience. 2016 July; 28(7): 1039–1051.

Heinrichs-Graham, Elizabeth, Wilson, Tony W. (2015). *Coding complexity in the human motor circuit.* Hum Brain Mapp. 2015 December; 36(12): 5155–5167.

Heinrichs-Graham, Elizabeth, Wilson, Tony W., Santamaria, Pamela M., Heithoff, Sheila K., Torres-Russotto, Diego, Hutter-Saunders, Jessica A.L., Estes, Katherine A., Meza, Jane L., Mosley R. L., Gendelman Howard E. (2014). *Neuromagnetic Evidence of Abnormal Movement-Related Beta Desynchronization in Parkinson's Disease.* Cerebral Cortex. October 2014; 24:2669–2678.

Heusser A. et al. (2016) *Episodic sequence memory is supported by a theta-gamma phase code.* Nat Neurosci. 2016 October; 19(10): 1374–1380.

Hipp JF, Engel AK, Siegel M. (2011). *Oscillatory synchronization in large-scale cortical networks predicts perception.* Neuron. 2011; 69:387–396.

Holmes G. (1918). *Disturbances of vision by cerebral lesions.* Br J Ophthalmol. 1918; 2(7):353–84.

Izhikevich, E. M. (2005). *Polychronization: Computation With Spikes.* Neural Computation June 15, 2005.

Izhikevich, E. M. (2007). *Dynamical Systems in Neuroscience: The Geometry of Excitability and Bursting .* The MIT Press, Cambridge, MA .

Izhikevich, EM, Edelman, GM. (2008). *Large-scale model of mammalian thalamocortical systems.* Proc Natl Acad Sci U S A. 2008 Mar 4;105(9):3593-8.

James, William (1890). *The principles of psychology.* New York: Holt.

Janet, Pierre (1899). *De l'Automatisme Psychologique.*

Jensen, O., Gelfand, J., Kounios, J., Lisman, J. (1999). *10-12 Hz oscillations increase with memory load in a short-term memory task.* NEUROIMAGE 9, 951-951.

Jerbi, Karim, Lachaux, Jean-Philippe, Diaye, Karim N, Pantazis, Dimitrios, Leahy, Richard M., Garnero, Line, Baillet, Sylvain (2007). *Coherent neural representation of hand speed in humans revealed by MEG imaging.* 7676–7681 PNAS May 1, 2007 vol. 104 no. 18.

Jia, X., Tanabe, S., and Kohn, A. (2013). *Gamma and the coordination of spiking activity in early visual cortex.* Neuron 77, 762-774.

Jones, E. (2002). *Thalamic circuitry and thalamocortical synchrony.* Philosophical Transactions of the Royal Society B. 357: 1659–1673.

Jones MW, Wilson MA. (2005). *Phase precession of medial prefrontal cortical activity relative to the hippocampal theta rhythm.* Hippocampus. 2005; 15:867–873.

Kim SM, Ganguli S, Frank LM. (2012). *Spatial Information Outflow from the Hippocampal Circuit: Distributed Spatial Coding and Phase Precession in the Subiculum.* The Journal of neuroscience. 2012; 32:11539–11558.

Khodagholy, Dion, Gelinas, Jennifer N., Buzsáki, György (2017). *Learning-enhanced coupling between ripple oscillations in association cortices and hippocampus.* Science 358 (2017): 369-372.

Klimesch, W., Doppelmayr, M., Schwaiger, J., Auinger, P., Winkler, T. (1999). *'Paradoxical' alpha synchronization in a memory task.* Brain Res Cogn Brain Res 7, 493-501.

Lago-Fernández LF, Huerta R, Corbacho F, Siguenza JA. (2000). *Fast response and temporal coherent oscillations in small-world networks.* Physical Review Letters 2000; 84:2758–61

Lampl, I., Reichova, I., Ferster, D. (1999). *Synchronous membrane potential fluctuations in neurons of the cat visual cortex.* Neuron 22, 361–374.

Lee U, Mashour GA, Kim S, Noh GJ, Choi BM (2009). *Propofol induction reduces the capacity for neural information integration: implications for the mechanism of consciousness and general anesthesia.* Consciousness and Cognition. 18 (1): 56–64.

Lee SY, Baftizadeh F, Campagnola L, Jarsky T, Koch C, Anastassiou CA. (2023). *Cell class-specific electric field entrainment of neural activity.* bioRxiv. Preprint. 2023 Feb 15:2023.02.14.528526.

Levine, J. (1983). *Materialism and Qualia: the Explanatory Gap.* Pacific Philosophical Quarterly. 1983. Vol. 64, № 4. P. 354—361.

Lewis LD, Weiner VS, Mukamel EA, Donoghue JA, Eskandar EN, Madsen JR, Anderson WS, Hochberg LR, Cash SS, Brown EN, et al. Rapid fragmentation of neuronal networks at the onset of propofol-induced unconsciousness. Proc Natl Acad Sci U S A. 2012; 109: E3377–E3386.

Liebe S, Hoerzer GM, Logothetis NK, Rainer G. (2012). *Theta coupling between V4 and prefrontal cortex predicts visual short-term memory performance.* Nat Neurosci. 2012; 15:456–462. S451–452.

Linkenkaer-Hansen, Klaus, Nikouline, Vadim V., Palva, J. Matias, Ilmoniemi, Risto J. (2001). *Long-Range Temporal Correlations and Scaling Behavior in Human Brain Oscillations.* The Journal of Neuroscience, February 15, 2001, 21(4):1370–1377.

Lisman, John E., Jensen, Ole (2013). *The Theta-Gamma Neural Code.* Neuron 77, March 20, 2013.

Lister WT, Holmes G. (1916). *Disturbances of vision from cerebral lesions, with special reference to the cortical representation of the macula.* Proc R Soc Med. 1916; 9(Sect Ophthalmol):57–96.

Liu Y, Liang M, Zhou Y, He Y, Hao Y, Song M, Yu C, Liu H, Liu Z, Jiang T. *Disrupted small-world networks in schizophrenia.* Brain. 2008 Apr;131(Pt 4):945-61.

Llinas, R. (2001). *I of the Vortex: From Neurons to Self.* The MIT Press.

Llinas, R., Ribary, U., Contreras, D., Pedroarena, C. (1998). *The neuronal basis for consciousness.* Phil. Trans. R. Soc. Lond. B 353, 1841–1849.

Llinas, R. R., Ribary, U., Jeanmonod, D., Kronberg, E., Mitra, P. P. (1999). *Thalamocortical dysrhythmia: a neurological and neuropsychiatric syndrome characterized by magnetoencephalography.* Proc. Natl. Acad. Sci. U.S.A. 96, 15222–15227.

Malhotra, S., Cross, R.A., Meer, M.A. (2012). *Theta phase precession beyond the hippocampus.* Reviews in the neurosciences. 23 1, 39-65.

Maris, E., van Vugt, M., Kahana, M. (2011). *Spatially distributed patterns of oscillatory coupling between high-frequency amplitudes and low-frequency phases in human iEEG.* Neuroimage 54, 836–850.

Mack, A., Rock, I. (1998). *Inattentional Blindness*, MIT Press

Marshall, Lisa, Helgadóttir, Halla, Mölle, Matthias, Born, Jan (2006). *Boosting Slow Oscillations During Sleep Potentiates Memory.* Nature 444: 610–613.

Marti S, Sigman M, Dehaene S. (2012). *A shared cortical bottleneck underlying Attentional Blink and Psychological Refractory Period.* Neuroimage. 2012; 59:2883–2898.

McKenna, T.M., McMullen, T. A., Shlesinger, M. F. (1994). *The brain as a dynamic physical system.* Neuroscience 60(3):587-605. June 1994.

Meshberger, F. L. (October 10, 1990). *An Interpretation Of Michelangelo's Creation Of Adam Based On Neuroanatomy.* JAMA Vol. 264, No.14.

Micheloyannis S, Pachou E, Stam CJ, Breakspear M, Bitsios P, Vourkas M, Erimaki S, Zerakis M. *Small-world networks and disturbed functional connectivity in schizophrenia.* Schizophrenia Research 2006; 87:60–6.

Mizuseki K, Sirota A, Pastalkova E, Buzsáki G. (2009). *Theta Oscillations Provide Temporal Windows for Local Circuit Computation in the Entorhinal-Hippocampal Loop.* Neuron. 2009; 64:267–280.

Montgomery, S.M., Buzsáki, G. (2007). *Gamma oscillations dynamically couple hippocampal CA3 and CA1 regions during memory task performance.* Proc. Natl. Acad. Sci. USA 104, 14495–14500.

Moruzzi, G., Magoun, H. W. (1949). *Brain stem reticular formation and activation of the EEG.* Electroencephalogr. Clin. Neurophysiol. 1, 455–473.

Nakazono T, Takahashi S, Sakurai Y. (2019). *Enhanced Theta and High-Gamma Coupling during Late Stage of Rule Switching Task in Rat Hippocampus.* Neuroscience. 2019 Aug 1; 412: 216-232.

Noback CR, Strominger NL, Demarest RJ, Ruggiero DA (2005). *The Human Nervous System: Structure and Function (Sixth ed.).* Totowa, NJ: Humana Press.

Palva S, Monto S, Palva JM. (2010). *Graph properties of synchronized cortical networks during visual working memory maintenance.* Neuroimage. 2010; 49:3257–3268.

Panksepp, J., Biven, L. (2012). *The Archaeology of Mind: Neuroevolutionary Origins of Human Emotion.* . New York: W. W. Norton & Company.

Pape, Anna-Antonia, Siegel, Markus. (2016). *Motor cortex activity predicts response alternation during sensorimotor decisions.* NATURE COMMUNICATIONS 7 Oct 2016.

Park, H., Kang, E., Kang, H., Kim, J.S., Jensen, O., Chung, C.K., Lee, D.S. (2011). *Cross- frequency power correlations reveal the right superior temporal gyrus as a hub region during working memory maintenance.* Brain Connect 1, 460-472.

Pashler, H. (1984). *Processing stages in overlapping tasks: evidence for a central bottleneck.* J. Exp. Psychol. Hum. Percept. Perform. 10, 358–377

Pautz, Adam (2019). *What is Integrated Information Theory?: A Catalogue of Questions.* Journal of Consciousness Studies. 26 (1): 188–215.

Pedroarena, C. M., Llinas, R. (1997). *Dendritic calcium conductances generate high-frequency oscillation in thalamo-cortical neurons.* Proc. Natl Acad. Sci. USA 94, 724–728.

Pedroarena, C. M., Llinas, R. (2001). *Interactions of synaptic and intrinsic electroresponsiveness determine corticothalamic activation dynamics.* Thalam. Rel. Syst. 1, 3–14.

Penttonen, M., Buzsáki, G. (2003). *Natural logarithmic relationship between brain oscillators.* Thalamus & Related Systems 2.

Perelman, J. (1967). *Entertaining algebra.* Moscow: Science.

Phillips JM, Vinck M, Everling S, Womelsdorf T. (2014). *A long-range fronto-parietal 5- to 10-Hz network predicts "top-down" controlled guidance in a task-switch paradigm.* Cereb Cortex. 2014 Aug;24(8):1996-2008.

Pitts W., McCulloch, W.S. (1947). *How we know universals: the perception of auditory and visual forms.* Bull. Math. Biophys V.9, 127—147.

Pizzella, Vittorio, Marzetti, Laura, Della Penna, Stefania, de Pasquale, Francesco, Zappasodi, Filippo, Romani, Gian Luca (2014). *Magnetoencephalography in the study of brain dynamics.* Functional Neurology 29(4): 241-25.

Purves, Dale, Augustine, George, Fitzpatrick, David, Hall, William, Lamantia, Anthony-Samuel, White, Leonard (2012). *Neuroscience. 5th Edition.* Sinauer Associates, Inc.: Sunderland, MA. ISBN: (Hardcover) 978-0878936953.

Rao, R. P. (2015). *Computational NS Section 1 Introduction part 4 The Electrical Personality of Neurons 17.30 min.* Coursera lectures.

Rao, R. P. N., Ballard, D. H. (1999). *Predictive coding in the visual cortex: a functional interpretation of some extra-classical receptive-field effects.* Nature Neuroscience, 2(1):79–87.

Rieke, F., Bialek, W., Warland, D., de Ruyter van Steveninck, R. R. (1999). *Spikes: Exploring the neural code.* MIT Press, Cambridge.

Robbe, D., Montgomery, S.M., Thome, A., Rueda-Orozco, P.E., McNaughton, B.L., Buzsáki, G. (2006). *Cannabinoids reveal importance of spike timing coordination in hippocampal function.* Nat. Neurosci. 9, 1526–15.

Ronnqvist, Kim C., McAllister, Craig J., Woodhall, Gavin L., Stanford Ian M., Hall, Stephen D. (2013). *A multimodal perspective on the composition of cortical oscillations.* Frontiers in Human Neuroscience April 2013 Volume 7 Article 132.

Rulkov, N. F., Sushchik, M., Tsimring, L.S., Abarbanel H. D. I. . (1995). *Generalized synchronization of chaos in directionally coupled chaotic systems.* Phys. Rev. E, 51(2):980–994.

Rulkov, Nikolai F., Bazhenov, Maxim (2008). *Oscillations and Synchrony in Large-scale Cortical Network Models.* J Biol Phys 34:279–299.

Ryun, Seokyun, Kim, June Sic, Lee, Sang Hun, Jeong, Sehyoon, Kim, Sung-Phil, Chung, Chun, Kee (2014). *Movement Type Prediction before Its Onset Using Signals from Prefrontal Area: An Electrocorticography Study.* Hindawi Publishing Corporation BioMed Research International Volume 2014.

Sarnthein J, Petsche H, Rappelsberger P, Shaw GL, von Stein A. (1998). *Synchronization between prefrontal and posterior association cortex during human working memory.* Proc Natl Acad Sci U S A. 1998; 95:7092–7096.

Sauseng P, Klimesch W, Schabus M, Doppelmayr M. (2005). *Fronto-parietal EEG coherence in theta and upper alpha reflect central executive functions of working memory.* Int J Psychophysiol. 2005; 57:97–103.

Schack B, Klimesch W, Sauseng P. (2005). *Phase synchronization between theta and upper alpha oscillations in a working memory task.* Int J Psychophysiol. 2005; 57:105–114.

Scheeringa, R., Petersson, K.M., Oostenveld, R., Norris, D.G., Hagoort, P., Bastiaansen, M.C. (2009). *Trial-by-trial coupling between EEG and BOLD identifies networks related to alpha and theta EEG power increases during working memory maintenance.* NEUROIMAGE 44, 1224-1238.

Schiff, S.J., Jerger, K., Duong, D.H., Chang, T., Spano, M.L., Ditto, W.L. (1994). *Controlling chaos in the brain.* Nature 370(August 25, 1994):615-20.

Schnitzler, Alfons, Gross, Joachim. (2005). *Normal and pathological oscillatory communication in the brain.* Nature reviews. Neuroscience Volume 6. April 2005.

Schoffelen JM, Oostenveld R, Fries P. (2005). *Neuronal coherence as a mechanism of effective corticospinal interaction.* Science. 2005; 308:111–113.

Schrödinger, E. (1944). *What Is Life?: The Physical Aspect of the Living Cell.* Based on lectures delivered under the auspices of the Dublin Institute for Advanced Studies at Trinity College, Dublin, in February 1943.

Schroter MS, Spoormaker VI, Schorer A, Wohlschlager A, Czisch M, Kochs EF, Zimmer C, Hemmer B, Schneider G, Jordan D, et al. (2012). *Spatiotemporal reconfiguration of large-scale brain functional networks during propofol-induced loss of consciousness.* J Neurosci. 2012; 32:12832– 12840.

Senkowski, Daniel, Schneider, Till R., Foxe, John J., Engel Andreas K. (2008). *Crossmodal binding through neural coherence: implications for multisensory processing.* Trends in Neurosciences Vol.31 No.8, 2 July 2008.

Sergent, C. et al. (2005). *Timing of the brain events underlying access to consciousness during the attentional blink.* Nat. Neurosci. 8, 1391–1400

Shadlen, Michael N., Newsome, William T. (1994). *Noise, neural codes and cortical organization.* Current opinion in Neurobiology. 4:569-579.

Shadlen, MN, Movshon, JA. (1999). *Synchrony Unbound: Review A Critical Evaluation of the Temporal Binding Hypothesis.* Neuron. 24 (1): 67–77, 111–25

Sherman, S.M., Guillery, R.W., 2002. The role of the thalamus in the flow of information to the cortex. Phil. Trans. R. Soc. Lond. B. 357, 1695-708.

Shirvalkar, P.R., Rapp, P.R., Shapiro, M.L. (2010). *Bidirectional changes to hippocampal theta-gamma comodulation predict memory for recent spatial episodes.* Proc. Natl. Acad. Sci. USA 107, 7054–7059.

Siebenhühner F, Wang SH, Arnulfo G, Lampinen A, Nobili L, Palva JM, Palva S. (2020). *Genuine cross-frequency coupling networks in human resting-state electrophysiological recordings.* PLoS Biol. 2020 May 6;18(5):e3000685.

Siegel M, Donner TH, Oostenveld R, Fries P, Engel AK. (2008). *Neuronal synchronization along the dorsal visual pathway reflects the focus of spatial attention.* Neuron. 2008; 60:709–719.

Siegel, M., Warden, M.R., Miller, E.K. (2009). *Phase-dependent neuronal coding of objects in short-term memory.* Proc. Natl. Acad. Sci. USA 106, 21341–21346.

Siegel, Markus, Donner, Tobias H., Engel, Andreas K. (2012). *Spectral fingerprints of large-scale neuronal interactions.* Nature reviews. Neuroscience Volume 13. February 2012.

Simonov, P. (1981). *The emotional brain.* Moscow: Science.

Singer, W. (1993). *Synchronization of cortical activity and its putative role in information processing and learning.* Annu. Rev. Physiol. 55, 349–374.

Singer, W. (1999). *Neuronal Synchrony: A Versatile Code Review for the Definition of Relations?* Neuron, Vol. 24, 49–65, September, 1999.

Sirota, Anton, Montgomery, Sean, Fujisawa, Shigeyoshi, Isomura, Yoshikazu, Zugaro, Michael, Buzsaki, Gyorgy (2008). *Entrainment of Neocortical Neurons and Gamma Oscillations by the Hippocampal Theta Rhythm.* Neuron 60, 683–697, November 26, 2008.

Sohal, V.S., Zhang, F., Yizhar, O., Deisseroth, K. (2009). *Parvalbumin neurons and gamma rhythms enhance cortical circuit performance.* Nature 459, 698–702.

Spitzer, B., Wacker, E., Blankenburg, F. (2010). *Oscillatory correlates of vibrotactile frequency processing in human working memory.* J Neurosci 30, 4496-4502.

Spratling, M. W. (2015). *A Review of Predictive Coding Algorithms.* Brain and Cognition 11.003.

Stam CJ, Jones BF, Nolte G, Breakspear M, Scheltans P. *Small-world networks and functional connectivity in Alzheimer's disease.* Cerebral Cortex 2007; 17:92–9.

Steriade, M. (1996). *Arousal: revisiting the reticular activating system.* Science 272, 225–226. doi: 10.1126/science.272.5259.225.

Steriade, M., Amzica, F. (1996). *Intracortical and corticothalamic coherency of fast spontaneous oscillations.* Proc. Natl. Acad. Sci. U.S.A. 93, 2533–2538.

Strogatz, S. (2003). *Sync: The Emerging Science of Spontaneous Order.* Hyperion.

Tallon-Baudry, C., O. Bertrand, C. Delpuech and J. Pernier. 1996. *Stimulus specificity of phase-locked and non-phase- locked 40 Hz visual responses in human.* J. Neurosci. 16: 4240–4249.

Tallon-Baudry, C., O. Bertrand, C. Delpuech and J. Pernier. 1997. *Oscillatory gamma-band (30–70 Hz) activity induced by a visual search task in humans.* J. Neurosci. 17: 722–734.

Tallon-Baudry, C., O. Bertrand, F. Peronnet and J. Pernier. 1998. *Induced gamma-band activity during the delay of a visual short-term memory task in humans.* J. Neurosci. 18: 4244–4254.

Tanigawa H, Majima K, Takei R, Kawasaki K, Sawahata H, Nakahara K, Iijima A, Suzuki T, Kamitani Y, Hasegawa I. (2022). *Decoding distributed oscillatory signals driven by memory and perception in the prefrontal cortex.* Cell Rep. 2022 Apr 12;39(2):110676.

Tegmark, Max (2016). *Improved Measures of Integrated Information.* PLOS Computational Biology. 12 (11): e1005123.

Thiele, A.; Stoner, G. (2003). *Neuronal synchrony does not correlate with motion coherence in cortical area MT.* Nature, 421 (6921): 366–370, Bibcode: 2003 Natur. 421. 366T

Thorpe, S. J. (1990). *Spike arrival times: A highly efficient coding scheme for neural networks.* In R. Eckmiller, G. Hartmann & G. Hauske (Eds.) Parallel processing in neural systems and computers (pp. 91-94): North-Holland Elsevie.

Tort, A.B., Komorowski, R.W., Manns, J.R., Kopell, N.J., and Eichenbaum, H. (2009). *Theta-gamma coupling increases during the learning of item-context associations.* Proc. Natl. Acad. Sci. USA 106, 20942–20947.

Tuladhar, A.M., Ter Huurne, N., Schoffelen, J.M., Maris, E., Oostenveld, R., Jensen, O. (2007). *Parieto-occipital sources account for the increase in alpha activity with working memory load.* Hum Brain Mapp 28, 785-792.

Van Albada Sacha, Robinson Peter. (2013). *Relationships between Electroencephalographic Spectral Peaks Across Frequency Bands.* Frontiers in Human Neuroscience. VOLUME=7. 2013. DOI=10.3389/fnhum.2013.00056.

van der Meer MA, Redish AD. (2011). *Theta phase precession in rat ventral striatum links place and reward information.* J Neurosci. 2011; 31:2843–2854.

van Wijk, Bernadette C.M., Neumann, Wolf-Julian, Schneider, Gerd-Helge, Sander, Tilmann H., Litvak, Vladimir, Kühn, Andrea A. (2017). *Low-beta cortico-pallidal coherence decreases during movement and correlates with overall reaction time.* NeuroImage 159, 1–8.

Vertes, Petra E., Duke, Thomas (2010) *Effect of network topology on neuronal encoding based on spatiotemporal patterns of spikes.* HFSP Journal. Vol. 4, Nos. 3-4, June-August 2010, 153–163

Verzeano, M. (1972). *Pacemakers, Synchronization, and Epilepsy. In: Petsche H., Brazier M.A.B. (eds) Synchronization of EEG Activity in Epilepsies.* Springer, Vienna.

Victor, Jonathan D., Purpura, Keith P. (1996). *Nature and precision of temporal coding in visual cortex: a metric-space analysis.* Journal of Neurophysiology Vol. 76 No. 2 August 1996.

Volk D, Dubinin I, Myasnikova A, Gutkin B, Nikulin VV. (2018). *Generalized Cross-Frequency Decomposition: A Method for the Extraction of Neuronal Components Coupled at Different Frequencies.* Front Neuroinform. 2018 Oct 18;12:72.

von der Malsburg, C. (Sep 1999). *The what and why of binding: the modeler's perspective.* Neuron. 24 (1): 95–104, 111–25.

von Helmholtz, H. (1867). *Handbuch der physiologischen Optik.* Leipzig, Leopold Voss, 1867 translated as Treatise on Physiological Optics, 1925 by Optical Society of America.

von Stein A, Chiang C, König P. (2000). *Top-down processing mediated by interareal synchronization.* Proc Natl Acad Sci U S A. 2000; 97:14748–14753.

von Stein A, Sarnthein J. *Different frequencies for different scales of cortical integration: from local gamma to long range alpha-theta synchronization.* International Journal of Psychophysiology 2000; 38: 301-13.

Watson, Brendon O., Buzsáki, György (2015). *Sleep, memory and brain rhythms.* American Academy of Arts & Sciences.

Watts, Duncan J., Strogatz, Steven H. (1998). *Collective dynamics of small world networks.* Nature 393.

Wilson, Matthew A., McNaughton, Bruce L. (1994). *Reactivation of Hippocampal Ensemble Memories During Sleep.* Science vol. 265 29 July 1994.

Womelsdorf T, Schoffelen JM, Oostenveld R, Singer W, Desimone R, Engel AK, Fries P. (2007). *Modulation of neuronal interactions through neuronal synchronization.* Science. 2007; 316:1609–1612.

Zandvakili, A., and Kohn, A. (2015). *Coordinated Neuronal Activity Enhances Corticocortical Communication.* Neuron 87, 827-839.

Zeki, S. (1993) *A Vision of the Brain.* Blackwell

Zeki, S. (2003). *The disunity of consciousness.* Trends in Cognitive Sciences. 7(5): 214–218. doi:10.1016/s1364-6613(03)00081-0. ISSN 1364-6613.

Zheng, Chenguang, Bieri, Kevin, Wood, Hsiao, Yi-Tse, Colgin, Laura Lee (2016). *Spatial Sequence Coding Differs during Slow and Fast Gamma Rhythms in the Hippocampus.* Neuron 89, 398–408, January 20, 2016.

Books in the Series
Symphony of Matter and Mind

Part one

Music of Matter
Mechanism of Material Structures Formation

Part two

Theory of Energy Harmony
Mechanism of Fundamental Interactions

Part three

Music of Life
Physics and Technology of Living Matter

Part four

Algorithm of the Mind
Teleological Transduction Theory

Part five

Technologies of the Mind
The Brain as a High-Tech Device

Part six

Harmonies of the Mind
Physics and Physiology of Self

Part seven

Inner Universe
The Mind as Reality Modeling Process

Part Eight

Dissonances of the Mind
The Physics of Mental Disorders

About the Author

Stanislav Tregub

Independent researcher.

Research areas: physics, biophysics, neuroscience, psychology, psychiatry.

www.ingramcontent.com/pod-product-compliance
Lightning Source LLC
Chambersburg PA
CBHW082106220526
45472CB00009B/2068